No Naughty Cats

No Naughty Cats

The First Complete Guide to Intelligent Cat Training

Debra Pirotin, D.V.M.
and
Sherry Suib Cohen

1817

HARPER & ROW, PUBLISHERS, NEW YORK
Cambridge, Philadelphia, San Francisco, London
Mexico City, São Paulo, Singapore, Sydney

Grateful acknowledgment is made for permission to reprint:

"You know how cats are" by Diane White. Reprinted by permission of *The Boston Globe.*

Excerpt from *The Cat's Pajamas: A Charming and Clever Compendium of Feline Trivia* by Leonore Fleischer. Copyright © 1982 by Leonore Fleischer. Reprinted by permission of Harper & Row, Publishers, Inc.

Library of Congress Cataloging in Publication Data

Pirotin, Debra.
  No naughty cats.

  Includes index.
  1. Cats—Training. I. Cohen, Sherry Suib.
II. Title.
SF446.6.P57   1985   636.8   84-48188
ISBN 0-06-015438-1

85 86 87 88 RRD 10 9 8 7 6 5 4 3 2 1

Contents

Acknowledgments vii

A Note on Equal Time ix

**1 You Can So
Train Cats! 1**

Self-fulfilling Prophecies 2

**2 Training and Health:
They Go Together 11**

Is the Problem Behavioral or Medical? 11

Defecation out of the litter box 14 • Indiscriminate urination 15 • Biting or other lashing out 17 • Hiding or irrational shyness 18 • Depression or listlessness 19 • Aggressive behavior 22 • Incessant crying 24

*Training and Nutrition:
Diets for Problem Times 25*

Anemia and behavioral problems 27 • Constipation and behavioral problems 28 • Blocked cat and behavioral problems 29 • Orphaned kitten and behavioral problems 30 • Diets 28

3 Training Your Cat to Stop Doing Something 33

The Big Five Problem Personalities:

The Killer Cat 33 • The Uncouth Cat 53 • The Crybaby 67 • The Destroyer 72 • The Prima Donna 79

4 Training Your Cat to Start Doing Something 87

PAR for the Training Course 89 • How to Train Your Cat to:

Come When Called 92 • Stay and Stop 99 • Be an Indoor Cat 101 • Accept Medication and Ministering 102 • Walk on a Leash 105 • Use a Cat Flap 108 • Love His Carrier 109 • Love His Bath 111 • Love Private Bathroom Habits 114

Cat Workouts 115: Jump Up 115 • Sit Up 116 • Ring for Out 117 • Over 117 • Fetch 118

How to Train Your Cat to Resist Danger 119 • Tricks You Won't Learn in This Book 123 • Cats and Babies Together 123

5 Cat Communiqués: You Can't Train Your Cat if You Can't Talk to Him 127

Body Language

Tongue Talk 130 • Head Talk 132 • Tail Talk 132 • Ear Talk 133 • Eye Talk 134 • Paw Talk 135 • Posture Talk 135 • Nagging 136 • Nose Talk 137 • Whisker Talk 137

Sounds

Purring 137 • Hissing 138 • Yowling and Growling 138 • Meowing 138 • Chirring 138 • Mating Call 139 • Talking Back 139

6 A Potpourri of Cat Smarts 141

Your Cat's Psyche 141 • Your Cat's Doctor 143 • Your Cat When You're Away 145 • Your Cat (Choosing the Best Candidate for Training) 150 • You Know How Cats Are 153

Index 157

Acknowledgments

Thanks to . . .

Our families for their constant support.

Colleagues at the Feline Health Center for their wisdom and encouragement.

Connie Clausen, the best literary agent in the whole world.

Trish Turk, typist extraordinaire.

Most of all, our much-loved editor, Larry Ashmead, Top Cat in our book.

A Note On Equal Time

If your cat's name is Tabitha, you're going to feel uncomfortable reading a book that refers to all cats as "he."

If your cat's name is Arthur, you're not going to love a book that refers to cats generically as "she."

Forget "it." I have never really thought of any cat as "it"—except maybe for a moment—so you'll rarely see training advice for "it" in this book.

It seemed that the most fair approach was to give he and she equal time, and so you'll see both pronouns used arbitrarily throughout *No Naughty Cats*. Pick the one that's right for your cat.

1

You Can So Train Cats!

Tell most cat owners that you can teach them to train their cats and the response is a guffaw or a knowing smirk. "You cannot train a cat," said one friend to me, "because the cat doesn't give a damn." Wrong, wrong, wrong. The cat *does* give a damn.

Tell yet another cat owner that you can train a cat, teach it to come, stay, jump, and she'll nod in intelligent agreement. "Of course you can train cats," she may say. "My cat's whole reason for being is to show off."

The truth lies somewhere in between. You can train cats, but you can't train them as you'd train a dog, and you can't train them to be utterly consistent in their responses because, well—they're just different. Compromises are definitely in order when you're talking cat training.

Nevertheless, it's possible, desirable and totally natural to teach a cat to respond to direction. If you *believe* that it can be done, and you communicate this belief totally to your cat, *and you don't expect circus tricks* (cats are far too wise to be demeaned by circus tricks), you *can* train your cat.

First, understand a bit about self-fulfilling prophesies.

You get the cat you expect to get—just as you probably get the kid you expect to get. Anyone who buys the legend that cats are mysterious and cold will have a mysterious and cold cat, just as anyone who buys the theory that kids are cruel will probably find himself with a cruel kid in his family.

If you believe, in your heart of hearts, that the "born free" legend is sacrosanct and that you are trampling on your cat's wild and wonderful nature by, say, spaying it, or training it to come when called—you will have a cat that's better off being returned to the wild (where it will probably die because cats are not snow leopards at heart) instead of a perfectly happy, reasonable house pet.

Absolutely foolproof studies have shown that kids who are expected to be bright, aggressive and friendly turn out to be just that. Not *such* absolutely foolproof (but, in my mind, just as accurate) studies show that cats who are nurtured, trained with respect and intelligence, and physically handled with love turn out to be responsive, contented animals.

Self-fulfilling Prophecies

Cat owners learn from experience, and the reason why experiences vary so much in owners' relationships with their cats is because each owner starts from a fixed position, creating the cat she expects to own. Some expect their cats to be willful and insular, and others expect theirs to be trainable. Of course, cats also differ in personality and genes—just as people do—and some cats are intrinsically easier to train than others, but on the whole you can get what you expect: You get what you train your cat to be.

Let's talk about cats and cat owners in general. I've seen more cat owners than most, and I feel qualified to make a generalization on the subject: Cat owners, as a breed, seem to be a resoundingly intelligent lot.

Cat owners almost invariably want to know *why* their pets behave as they do. They seem to be respecters of independence and individuality, both of which are hallmarks of feline behavior. Cat owners usually have an acute sense of humor, more highly developed than, say, dog or parrot owners. You won't find *most* cat owners painting their cat's claws with nail polish, or shaping their cat's fur into little poodle-y poufs. The average cat owner is simply not interested in having her animal roll over and play dead—that seems dumb to her. Good thing . . . few cats will do it.

In short, cat owners deserve a down-to-earth, intelligent and unsimpering book that will not exhort them to take their cats to be psychoanalyzed or have their astrological signs read. They want a no-nonsense, pragmatic guide to teach their animals to be well behaved, pleasure giving and easy to live with. Since no cat owner I've ever known has been even remotely tempted to send a birthday card to his cat saying, "Happy Birthday, You Wonderful You," it's hard to understand why, as a rule, books about cats are written so cutesy poo. Cat owners, you may be sure, don't love reading articles about cats that spell perfect as *purr*-fect. If you spot purr-fect anywhere in this book's text, return it to the book store and get your money back.

Why, then, should the legions of cat books on the market for the most part gear their advice to people who seem to be of reduced I.Q.? Why are the books on the market patronizing and oddly childlike? It's a paradox. This is why I firmly believe that most *writers* of books on cats don't really understand their market—either the cats or the cat owners, and that's why so few maintain that training is possible and desirable. Result: cat owners, intelligent as they are, have been accepting the myth that says, "Don't even try to train your cat," when they should really know better.

Let me do what I can to strike down this myth. Cats, it is generally preached, are untrainable because they are still denizens of the wild—unlike the dog who has been domesticized from his wolf ancestors.

Cats are no more wild creatures or comparable to jungle cats than man is comparable to apes. It's been a long time since a cat was a lion. Cats were not dragged, kicking and screaming, from the jungle last week and forced to come live in houses. We are not denying their atavistic natures by training them. I speak from experience, not wishful thinking.

When I first decided to be a veterinarian, I was entranced by the exotic animals—in particular, the big cats. After earning an undergraduate degree in pharmaceutical medicine at Columbia University, I was drawn to a veterinarian school in the Philippines where I could learn about the natures of the wild animals in their natural environments. There, I had unique, firsthand experience. I treated a monkey-eating eagle and caribou along with wild cats and exotic birds. As a student, I entered the homes of the townspeople and helped treat their cherished pets; in this country, I would only have been permitted to observe the doctor in charge of a veterinarian clinic. I had hands-on, immediate experience, along with the book-learning and it was, I think, a wonderful way to become a caring doctor. As I studied the exotic and domestic animals of the Philippine Islands, my own six cats (Rocky, Abby, Sneakers, and The Three

Un-Named) came and went at will. The more I worked with cats, the more I knew they would be my specialty. I've always had a love affair going with them. To complete my education, I was granted two preceptorships: one at the Animal Medical Center, in New York, and one at the Bronx Zoo. It was at this zoo and at the Manila Zoo from where I'd just come that I actually began to understand the differences and the similarities between domestic cats and the bigger, wild ones.

The similarities? As I made my morning rounds to care for the snow leopard, the lioness and the tigress, I was infinitely saddened to see their flaccid muscles, their lusterless coats and dull eyes, their friction sores from lying around in a confined area. Today, the Bronx Zoo, like many other large zoos, understands the care of wild animals far better and allows them to roam freely in large, fenced-off terrain, but then, those magnificent animals were ill with boredom, ill with confinement. Like inmates in mental institutions, they vacillated between angry aggressiveness and resigned, stifling despair. They were dying from lack of love, also, I believe.

Smaller cats, *our* cats, also need room to roam and they need interesting, stimulating environments—or else they too seem to die a little. If they're not caressed, talked to and played with, they become sour, unloving creatures. Like the big cats, they show *physical* symptoms of withering away if they are not treated with affection and attention. The big cat, roaming free in a pack, has endless opportunities to interact with his peers, to play, protect, run free with them. He thrives on attention of one sort or another from others. The little domestic cat does also.

Further, the cats with whom we live *learn* in similar ways to jungle cats. You can't, for instance, hit them and expect compliance. If you beat their jungle brothers with a palm branch, those cats will still not learn a lesson—and their descendants similarly will not respond with obedience to a swat on the flanks, as, say, a dog might. When you swat a cat, it will almost always either run away or fight back with clawing or biting. It will not learn to do the thing you want it to do.

Because their response to humans is different, in many ways, from dogs, cats will also not respond *as* predictably or wholeheartedly to the incentives of nurturing or stroking as a dog will. If you're affectionate to a dog, it will probably be—just as legend has it—your best friend forever, and heaven help the robber who comes to knock the pins out from under you. A cat who's been loved and petted will learn better and be sweeter but will not grow up to be your loyal attack cat, leaping at intruders' throats, even if you're very nice to it. Well, it *might,* but you just can't *depend* on its being a guard cat.

Another similarity between domestic cats and wild cats is that you can't crowd any cat and take away his need for a territorial claim and still expect him to be sound in mind and body—and that was true for his great-great-grandfather, as well.

You must provide him with the atavistic necessities his ancestors had, like substrate in which to eliminate, and rough material on which to claw, if you hope to have him live in mutual comfort and civilization with you. In these small ways, you must simulate the atmosphere of the wild in your home. Actually, *because* the cat shows consistent responses to certain stimuli, it makes a better candidate for a few training techniques than your dog could ever be. For instance, a cat has a propensity to urinate and defecate in substrate (sand, clay, wood chips—whatever). This makes it a natural for training to eliminate in the toilet if you should so wish to train your cat (see page 65). You use his substrate as a training *tool.* Try training your dog to urinate on the toilet. Simply can't be done.

But, given the similarities to its ancestors, what makes a house cat *different* from a jungle cat?

Simple. House cats have, quite naturally, used the years to learn to live with people in perfect compatibility. They were living compatibly with the good people in India and Egypt almost five thousand years ago and by now they've gotten the hang of it. I would not be happy sleeping in the same house with an uncaged leopard even if she were Siamese cat size. Leopards have not gotten the hang of it, yet. What's more, today's cats (the small variety) have been *trained* successfully for thousands of years. The Egyptians trained cats to be hunters and the felines were especially successful in controlling the rat population that systematically destroyed Egyptian grains. In China, cats were trained to guard the silkworm cocoons from other animal invaders. In Japan, cats were deified and venerated even after their deaths because of their roles as temple guardians. Siamese warriors used to train cats to sit on top of their shoulders and screech out cries of warning if an enemy approached from behind.

Now, larger jungle cats have been trained to do circus acrobatics but I've never heard of a wild cat who was trained to really live easily with humans, freely giving as much as it received. That's a very big difference.

Exotic cats—the lions, leopards and tigers of the world—must, in my opinion, live free to be content. I've never seen a leopard who, given his choice, wouldn't elect the jungle over the living room or the circus tent. But *our* cats? Not so. Given their choice, they almost always return home. They know a good thing when they see it.

Sure, you can't unreasonably manipulate cats. That's precisely why they're terrific ... and the intelligent person is drawn to them. There is fascination in an animal who lives sweetly with you and yet is not entirely predictable, is always subtle and restrained. With a dog, on the other hand, "what you see is what you get." When dealing with cats, you get a lot more than what you see.

And so, house cats are not wild creatures whom we insult by spaying, training and domesticating, but they are also not wet clay—totally moldable. Not mysterious or cold like Kipling's "Cat That Walked By Himself," they are in many ways like your own kids—shapeable up to a point, reasonable, reactive ... and still, quite individual. Without anthropomorphizing cats, I can easily say that if you raised a cat by reading Dr. Spock, you'd probably have a terrific kid—I mean, cat—on your hands.

An example: If you hardly ever touched your baby, and you hardly ever cooed and sang to her, your baby would grow up stunted and needy and hostile—even if you fed her the most balanced meals in the world. Medical evidence has shown that babies actually develop marasmus (from the Greek for wasting away) and become listless, withdrawn, slow in mental development or even fatally ill when they are not cuddled, touched or held. If you were a visitor from another planet, and you saw such a baby, you might think that all earthling babies are programmed to be sickly and mentally slow.

Similarly, if you never cuddle your cat, croon to it, love it with sweet and soft language, you will end up with a cold, withdrawn, quite neurotic cat.

If you never teach your cat to come when called, if you never *expect* it to live reasonably with you and your friends and family, it won't. A baby, unloved, with few expectations of its capacity, will grow into a sullen, sad and solitary creature. So will your cat.

Both may forget or choose to ignore the training and love you've given on certain occasions, but, for the most part, kids *and* cats respond to loving, intelligent handling and training . . . of course, to the degree each is capable of.

The other day I heard of Rhubarb, the seeing-eye cat. Her mistress, a Mrs. Elsie Schneider from California, lost all vision in an automobile accident. No one told Mrs. Schneider that cats cannot be trained and so she taught Rhubarb to guide her to the trash can, the mail box and the laundromat. Rhubarb learned to emit a special meow when they arrived at a destination, and to wait at curbs till traffic passed and she could safely guide Mrs. Schneider across. The cat recognized the names of neighbors and would take Mrs. Schneider to their homes when commanded. She was an extraordinary cat, granted, and her tombstone is in Braille in respectful tribute, but she was probably no more or less intelligent than *your* cat. She was the heroine of a self-fulfilling prophecy, a prophecy that Mrs. Schneider created when she decided she *would* have a seeing-eye cat. And so she did.

In short, it is natural to train a cat to live comfortably with people and to learn certain habits that will provide greater security *and* freedom for him. Your cat is not a snow leopard at heart—he will be infinitely happier with a happy family, and thus must learn the parameters of civilized pet/human bonds. Any street cat, any Burmese or Siamese is worlds apart from the big jungle cats who once were immediate family. I know from experience.

Your cat is eminently trainable, and you are not taking away a whit of its precious, natural instincts by teaching it companionable habits. In fact, you are adding immeasurably to its talents—and your mutual cohabitation.

Furthermore, you will outwit the newest trend—the growing legion of cat psychologists who, for upward of fifty dollars an hour, will help you to work out your cat's "relational problems"; in the end, all they really do is tell you how to modify your own behavior so as to stop bothering the cat they claim can't be trained. They're wrong. Train your cat. It *can* be done!

2

Training and Health: They Go Together

Is the Problem Behavioral or Medical?

I can't tell you how many behavioral problems are not simply misbehavior but are caused by the cat's not feeling well. In all my years of practice, it has always shocked and saddened me to see the large numbers of cats who are put to sleep or otherwise labeled incorrigible by owners who have not dug deeply enough into the physical causes of their misbehavior. So often, obnoxious feline behavior can stem from medical problems which, treated medically and not behaviorally, magically clear up. A cranky "Garfield" cat turns into a cupcake when his owner understands that he's cranky because his stomach hurts.

Into my office today came Fat Cat, Silver and Butterscotch, all of whom might have been diagnosed as behavioral pests had their owners not been as discerning as they are.

Fat Cat is a Coast Guard cat, a kind of mascot who lives on a little island adjacent to lower Manhattan with the men of a United States Coast Guard contingent that is based there. He hangs out with the guys, when he feels like it, and in nice weather sleeps in secret places on the island. For the last two weeks, says the burly, tough (but inside, tenderhearted) coastguardsman who brings him to me, Fat Cat has been crying without stop, waking up the night watch and, in general, making a whole lot of fuss. At first, the guys ignored the cat, thinking he was making a pitch for snacks, but finally, the crying became very annoying and two of the men were assigned (unofficially) to present Fat Cat to a doctor for diagnosis. He was tenderly carried into my examining room, a perfect Coast Guard cat—big and tough, one yellow, one blue eye—and mewing pitifully. Examination disclosed that Fat Cat was blocked with urinary calculi— a condition known as FUS (feline urologic syndrome) which, simply put, means that the urinary passage was occluded with plugs of tiny stones, mucus, blood and other organic bits. He hurt. So did the United States Coast Guard—for him. If they'd ignored Fat Cat's vocal complaints and chalked them up to pestiness, the cat soon would have died because FUS can result in fatal uremic poisoning in a very short time. I'm happy to say that Fat Cat was unblocked successfully and returned to guard New York against sea invasion, along with his pals.

Silver arrived in my office nipping and biting her owner's hand—a very new owner, it turned out, since Silver had just showed up on his stoop a few days before. The cat was silvery beautiful and something about her was irresistibly appealing to her newly adopted person, but he couldn't figure out her bad temper. An examination showed enormous infestation of ear mites, an abcessed wound (perhaps from a fight), not to mention an absolute flea invasion on her body. No wonder she was testy. We suggested isolating her in my office for a while, just to see if she was incubating a serious infection, before her owner introduced her to his other cats; after a week Silver was sent home, cured and even-tempered.

Butterscotch was the third cat brought in today with a "behavioral" problem. The day before, she was caught licking her chops from the tasty garbage. She seemed to be surly and snippy to her owners after five years of being the world's most perfect cat. What's even worse, she'd begun to urinate indiscriminately all over the place. Her owners tried giving her more affection and less water, but Butterscotch seemed frantic for more, not less liquid. A simple blood test told me what I'd already expected: Butterscotch had diabetes, the symptoms of which are similar to diabetes in humans—excessive thirst and urination, weight loss, odd moods. She was placed on a strict diet (the garbage binging had caused dangerous, high blood sugar) and insulin injections. In a week, she was the same old, terrific Butterscotch, much to her owners' relief.

So misbehavior can and often is prompted by physical problems. Cat owners have a responsibility to see that their cats are healthy before they attempt to train away their misdeeds.

Table 1, which follows (see page 14), is not by any means complete, because a whole book could easily be devoted to a truly comprehensive table. Still, it is a start to help you determine if the most common complaints of feline misbehavior should really be attributed to organic problems. This is the way to use the table: First, identify your cat's PROBLEM. Determine if it's medically induced by reading the CLUES to see if your cat is behaving in a way that clearly points to pain or discomfort instead of anger or stubbornness. Check out WHAT YOU CAN TRY to see if there are any home remedies that might

Table 1
Is the "Behavioral" Problem Really a
Medical Problem?

Problem: Defecation out of the litter box

Possible medical causes	Blocked anal glands, diarrhea, or not a medical problem but merely an accident—feces stick to tail area and then fall off, simulating indiscriminate defecation. Stress also can cause gastrointestinal disorders and/or diarrhea.
Clues: accident	Check to see if the cat has "stud tail," a condition caused by overactive sebaceous glands along the top of the cat's tail, near the base. If so, the oily sebum tends to trap feces during a proper, in-the-litter box elimination, and the feces drop off during the cat's daily activities.
Blocked anal glands	The cat may scoot along the floor on its rump to relieve pain or itch of blocked glands. It may also constantly lick the area, or cry in pain.
Diarrhea	May be caused by illness, improper diet, eating foreign objects or spoiled food, or even hairballs. Movements are excessively frequent and loose or watery.
What you can try	Clean up that tail! (if it's scruffy). Clip the hair, dust with cornstarch, which will absorb oil, or ask your vet for the best remedy. If it's blocked anal glands, you *must* see your doctor and not experiment with home remedies: the result could be disastrous if the cat's not unblocked soon. If it's diarrhea, try limiting the cat's diet to very bland foods like boiled chicken and rice or using any antidiarrheal medication that contains the substances called pectin and kaolin (sold over-the-counter in pet stores). Or try a small amount of milk of bismuth or Kaopectate. If the condition persists for 48 hours, see a doctor.

Table 1
Is the "Behavioral" Problem Really a
Medical Problem? (*continued*)

Medical solutions	Hairballs can be removed, diets changed and cats unblocked by draining the abcess, or plug, causing the blockage. Occasionally, tumors cause anal sac blockage and, in this case, the entire anal sac can be removed. Fecal analysis is required if persistent diarrhea is present to determine if diet or parasites is at fault. *Note:* Social stress (a poorly socialized cat, orphan kitten or runt of the litter who is being abused by other cats or animals) can lower the immunologic potential of the cat and gastrointestinal disorders can result. These disorders can cause indiscriminate defecation and flatulence.

Problem: Indiscriminate urination

Possible medical cause	Bladder infection (cystitis), diabetes, urinary incontinence from old age, arthritis (if the cat feels back pain, for example, it may strain to relieve that arthritic pain or muscle spasm and lose control of the bladder).
Clues	•Increased frequency of urination. •Cat may tend to urinate on cool surfaces (tub, tile floor) because it relieves pain. •Decreased volume of urine. •Enlarged abdomen. •Presence of blood in litter box (bright red to light pink). •Lethargy. •Obvious distress when urinating—straining, returning to litter box too often. •Increased thirst. •Crying.

Table 1
Is the "Behavioral" Problem Really a
Medical Problem? (*continued*)

What you can try

- If you suspect cystitis (increased frequency of urination is the biggest symptom), try a vitamin C tablet of 250 milligrams once a day until your vet can see you. It's possible the problem *may* clear up by itself in a day or so.
- *But*, if you suspect that your cat is **blocked** (he/ she is straining to urinate and is crying in pain), get to the veterinarian *immediately*. This is a life-or-death emergency situation and the cat must be treated immediately. Blocked cats may suffer for two or three days and then die because their owners procrastinated in getting help. Cats may die in the space of a day if uremic poisoning sets in.
- Sometimes a bladder infection can come from an improper dietary balance, so make sure your cat is getting the proper nutrients in a balanced diet— and is eating a *moist* cat food, rather than dry. Drinking liquids often helps to cure cystitis; you can try salting the cat's food to encourage drinking water.
- Weight loss sometimes helps arthritis: put your cat on a diet!

Note: Sometimes it's difficult to spot blood droplets in the urine: try a little shredded paper on top of the kitty litter. The blood will become apparent, if it's there, and tell you that you definitely have a medical, not a behavioral, problem. The shredded paper will also tell you if any urine is being produced, before it evaporates from the regular litter (a cessation or a decreased volume of urine indicates a medical problem, also).

Table 1
Is the "Behavioral" Problem Really a
Medical Problem? (*continued*)

Medical solutions	If your cat is urinating very much, drinking copious amounts of fluid and is generally lethargic or acting erratic, diabetes or renal disease may be a sure diagnosis (urine and blood tests will tell). If diabetes is present, insulin injections (just as people are treated) are a possible solution. If the problem is a secondary infection, antibiotics may be used to clear it up. Bladder stones may have to be removed, if they're present. Sometimes the urine is made more acidic with medication to prevent obstructions from forming but, if they're already there, catheterization is performed to relieve a blocked urinary tract. A low-magnesium diet will then be prescribed to stop the production of crystals.

Problem: Biting or other lashing out when being handled

Possible medical causes	Your cat may have a wound on its body or the problem may be as simple as static electricity, which gives your cat a shock whenever you touch it. Tumors can also cause pain when touched.
Clues	•Cat shies away whenever certain part of body is touched. •Cat's hair literally "stands on end" or fluffs out, unexpectedly (static electricity). •The cat may cry, lick or avoid using the area with the wound or injury.

Table 1
Is the "Behavioral" Problem Really a
Medical Problem? (*continued*)

What you can try   "Ground" yourself before touching cat (if you suspect static electricity) by touching a non-metallic substance like wood. Or, dampen the cat's fur with a wet cloth before you groom him if you suspect that static electricity may be present. Carefully inspect your cat's body, when it's drowsy and resting contentedly. Slowly run your hands over the length of the cat, looking for a wound, a bump or a bulge. Medically, this is called palpation and every owner should regularly palpate her cat. Superficial wounds can be treated by thorough cleansing with soap and water or peroxide.

Medical solutions   Your vet will examine the cat thoroughly and treat abcessed or infected wounds, or growths. Many cancerous tumors can also be treated today, with surgical excision and/or chemotherapy.

Problem: Hiding or irrational shyness

Possible medical causes   Cat may have a temperature or be otherwise ill with anemia or other diseases. Cat may be having a shock reaction after a fight or a fall or hearing a terrifying noise.

Clues   •Cat disappears for long periods of time into secluded boxes or crawl spaces.
•Cat's nose is particularly dry, warmish and cracked, possibly around the edges.
•Cat's eyes appear glazed.
•Cat has bad breath or drooling, which may be a sign of illness.
•Cat stops eating normally.
•Cat shivers and sweats from footpads.

Table 1
Is the "Behavioral" Problem Really a
Medical Problem? (*continued*)

What you can try

Take the cat's temperature. Use a rectal thermometer (a heavy-duty one), lubricate it with Vaseline. Get a friend to hold the cat's front end steady as you insert the thermometer into the rectum very gently and slowly until it's halfway in. After holding the cat and the thermometer steady for about 3 minutes, gently remove it. The cat's healthy temperature should be about 101° to 101.5° F and can vary a half a degree or so either way and still remain normal. If your cat's temperature is over 103, see a veterinarian as soon as possible. How high is *terrible?* Put it this way: Brain tissue in a cat dies at about 109° F. Gentle treatment and soothing voices go a long way to eliminating a shock response.

Medical solutions

The variety of illnesses for which a cat can run a temperature are many. Proper diagnosis will determine proper medical procedure. Do not ever give aspirin to your cat without a doctor's prescription. In many animals aspirin treatment can be lethal, despite safe human consumption, and cats react very strongly to this drug; it has to be given sparingly and rarely. Tylenol is *always* lethal to a cat.

Problem: Depression or listlessness

Possible medical causes

Anemia, removal of a kitten or a loved one (separation), increased temperature, overweight, heart disease, asthma, or hypothyroidism.

Table 1
Is the "Behavioral" Problem Really a
Medical Problem? (*continued*)

Clues	• Cat's too quiet, shy, sleepy (anemia?).
	• Cat may be hiding and have no appetite (if it has a temperature).
	• Cat may have pale nose, grayish gums and eye membranes (anemia symptoms).
	Note: The best way to determine if a cat's nose is paler than it should be is to compare it to another cat.
	• Cat may be overly warm to the touch (if it has a temperature).
	• Cay may cry (if it's orphaned or has lost its favorite person).
	• Cat may be overweight.
	• Shortness of breath.
	• Cat seems to crave heat.
What you can try	*If the cat is depressed because it's sad:* Good old TLC (tender, loving care) is the prescription.
	If the cat has a temperature: It's always better to find the root cause of the fever, rather than just treat the symptom. Still, if you want to get that fever down in a hurry while you're waiting to see the vet, a cold water paw and temple bath works wonderfully to bring the fever down. Plunging the whole cat into cold water can be a terrible shock to its system. *Never* give aspirin without a doctor's advice, *never* give Tylenol, period! and never give an alcohol rub for a fever. Cats are different from kids in this way.

Table 1
Is the "Behavioral" Problem Really a Medical Problem? (*continued*)

If you suspect your cat may be **anemic,** you can try the addition of a vitamin B complex and iron in its diet (anemic cats, remember, have very pale mucous membranes).

Medical solutions

In the case of extreme depression, doctors can try chemical mood elevators. Anemia may indicate the presence of other diseases and the cat with anemia must be treated as the *whole* cat. A microscopic parasite that destroys red blood cells or feline leukemia may well be the culprit.

Too often, owners ignore their cat's obvious weight problem as a possible cause for listlessness or depression. One of my clients with a penchant for pudding owned a 24-pound Persian cat that should have weighed less than half that. Every night, she and the cat sat together in front of the television set eating pudding—a spoonful for me, a spoonful for you, darling. Every night, each consumed a cup or more of pudding. The cat cried nightly and would no longer jump into the client's lap, which worried and troubled her. Was it arthritis? she wanted to know. Was it an orthopedic problem? No, it was pudding. She's lucky the cat didn't have heart problems as well, what with all that cholesterol circulating around in its blood. I put the cat on a diet and the patient also dropped about twenty pounds since she couldn't bear to eat pudding with the cat watching balefully with accusing eyes. No more meowing, listlessness or depression.

Table 1
Is the "Behavioral" Problem Really a
Medical Problem? (*continued*)

Important figures to remember: A cat that's overweight by 20 percent faces a mortality rate that's 50 percent greater than normal! If the cat's only 10 percent overweight, he still has a 33 percent greater chance of dying earlier than he would at a normal weight. Obesity is painful to pets, causes sluggishness, circulation problems and a higher risk of diabetes. Don't feed your cat nine treats for every successful learning experience, don't bring it home "kitty bags" filled with greasy stuff from restaurants, and don't feed it more than twice daily. If your cat has been in the habit of sitting around all day, watching television and eating bonbons, it's never too late to turn over a new leaf. Encourage it to jog by getting it its own cat: two are really no more trouble than one and are built-in exercisers for each other.

Problem: Aggressive behavior

Possible medical causes	The cat may have hormonal problems—may need to be castrated. He may have nutritional deficiencies. He may even be *over*-vitaminized (too many vitamins can be as lethal for cats as for people). The cat may have a thyroid problem.
Clues	If the problem's **hormonal:** •Cat sprays—all over! •Cat is *bursting* to escape: waits at door to bolt the instant a careless person leaves the door open. •Cat fights with other cats.

Table 1
Is the "Behavioral" Problem Really a Medical Problem? (*continued*)

•Cat may be skittish, hyperactive (hyperthyroid problem).
•May lose weight or have diarrhea (hyperthyroid problem).
If the problem's **nutritional:**
•Cat may be emaciated and have skin lesions or scaly and dry fur.
•Cat may have dull coat.
•Cat may experience lack of appetite.

What you can try

If the problem is nutritional, make sure the cat's diet is sound and complete. Do some reading about feline nutrition and consider a moist, commercially prepared cat food that has been approved by the NRC (National Research Council). If your vet has diagnosed specific nutritional problems, see page 31 for a home-prepared diet that may be just the thing for him.

Medical solutions

If your cat is aggressive because he needs neutering, have it done immediately! You'll have a much nicer and happier pet and you won't alter his or her essential catness, at all.

If your cat has a hyperthyroid problem, and the secretions of the thyroid gland are overabundant (this is diagnosed through a blood sample), your vet can treat the condition with radioactive isotope therapy (*not advised* if there are young children in the house), surgery or medication. Hypothyroid problems are treated chemically.

Table 1
Is the "Behavioral" Problem Really a
Medical Problem? (*continued*)

Problem: Incessant crying

Possible medical causes	Many physical problems that could be contributing to crying are too frequently diagnosed as a behavioral problem. Your cat may have a urinary blockage, arthritis, occluded anal sacs, constipation or pain from an accident or fight with another cat. Never assume crying is behavioral until you've ruled out illness or injury.
Clues	Palpate your cat (feel gently all over his body) and see if any area is more tender than others, indicating injury or even blockage. See pages 14 and 15 for "Defecation Out of the Litter Box" or "Indiscriminate Urination" for additional clues.
What you can try	If your cat is crying in a low, throaty voice or in a series of tiny yips for an extended period of time, and you cannot see any possible emotional reason for the crying (the cat hasn't lost a loved person or other cat, has not been left alone suddenly for long periods of time, and does not respond to loving, gentle attention—or calming heat), then you must assume that the crying is medically based. Your doctor's help is needed.
Medical solutions	Crying can be caused by so many things; an exhaustive examination and case history will help your doctor determine the cause and solution.

be useful, either permanently or even for a temporary cure in an emergency. Then, read the MEDICAL SOLUTIONS to see what your veterinarian will (or should) do to alleviate the condition. It's amazing how many naughty cats become good cats almost instantly, when a physical problem is creating behavioral annoyances. IF you can't see any possible medical problems behind the obnoxious behavior, *then* turn to chapter 3 and read "The Big Five Problem Personalities."

Training and Nutrition: Diets for Problem Times

There is no way you're going to have a well-trained cat if it is suffering from poor health, and feline health is in large part predicated on proper nutrition. It always amazes me how few clients realize that a cat's emotional/behavioral and physical well-being are directly related to nutrition. These clients are victims of brilliant media thrusts that lead them to believe that a cat loves tuna— therefore, tuna (or another fish product) is all he ever needs. Tuna, as a matter of fact, has a seduction value based mainly on smell and, indeed, a cat raised primarily on tuna almost invariably shows vitamin E deficiency problems—itchy skin, nervousness, aggressive traits and muscle disorders. Although this book is primarily concerned with training and is not a catchall of cat raising, which would include general nutritional advice (you can buy many such excellent *general* cat books), I think that training and behavioral problems are so related to certain nutritional problems that diet is important enough to bear discussing. Food requirements for cats are different from that of other animals—they require a diet high in protein and animal fat, and low in ash, oil-soluble vitamins (A, D, E and K), for instance. Serious nutritional overloads (with resulting behavioral problems) can come about from the indiscriminate feeding of vitamins.

I feel strongly that the best chance for cat health comes from feeding a cat a *commercial* and nutritionally sound, complete cat food (moist, not dry). They're tested again and again and remain the most all-around, no-guessing, no-human-error kind of diet. But many of my clients prefer holistic type foods (often for themselves as well as for their cats), and some cats certainly do have physical problems which create a need for natural home cooking, if not permanently, then at least temporarily. An excellent commercial holistic cat food is the Cornucopia-brand label. For clients who prefer to cook for their cats temporarily *or* permanently, I've gone to great lengths to find intelligently balanced recipes for home-cooked meals . . . and I've found them!

Doctors Mark L. Morris, Jr., and Lon D. Lewis of Mark Morris Associates in Topeka, Kansas, have made intensive study of the problem and have come up with the very best home recipes devised. They're included in a booklet called *Guide to Nutritional Management of Small Animals,* and if you call Mark Morris Associates in Topeka, you can order the free booklet. I've tried every recipe on various patients and have not been disappointed with their quality. The address is: Mark Morris Associates, 5500 Southwest Seventh, Topeka, Kansas 66606 (telephone 913-273-5055).

Thus, I have included in this book options for home-cooked meals for those cats who may exhibit behavioral or training problems—but who are, in reality, suffering from medical problems. If your cat has the symptoms listed in the next few pages, and you wish to cook for it until the problems clear up, here are some superior, balanced diets that Mark Morris Associates suggests you may safely try. You will find a recipe for home cooking for each problem. And, if you have neither the time nor inclination to cook at home for your cat, I also offer *the ingredients that should be included in a commercial food* that will best serve your cat and its particular ailment. These also are the recommendations of Mark Morris Associates. Coordinate your reading of this section with Table 1 (see pages 14–24).

1. If your cat has **anemia** (symptoms: depression, irrational shyness, pale mucous membranes, tiredness), try the diet below. (See Table 2, page 28, for ingredients of a commercial diet that would be suitable for anemia.)

Home-Cooked Meal

1 pound regular cooked ground beef
¼ pound liver (cooked)
1 cup cooked rice
1 teaspoon vegetable oil
1 teaspoon calcium carbonate
Balanced supplement (powder or liquid) which fulfills the feline minimum daily requirement for all vitamins and trace minerals

Combine all ingredients. Follow directions on the supplement you choose as to the amount. (We suggest Pet Tab or Felobits as possibilities.)

Yield: 1¾ pounds

Before beginning the diet check with your doctor and try the following until the hematocrit value approaches normal, or for a maximum of two weeks. (The hematocrit value is the result of a simple blood test to determine if your cat has anemia.)

• Increase protein level of diet by adding one part (by weight) muscle meat, liver, kidney, cottage cheese, or hard-boiled eggs to each three parts of customary food.

• Increase level of B-complex vitamins to six times minimum requirement. Supply at least 12 micrograms folic acid, 0.5 milligram niacin, 50 micrograms pyridoxine and 1.8 micrograms vitamin B_{12} per pound of body weight daily. These may be provided by adding 1 gram per pound per day of brewer's yeast to the diet.

• Increase iron, cobalt and copper intake with metallic salts by administering sufficient quantity of a trace mineral supplement to supply at least 3.5 milligrams iron, 0.5 milligrams copper and 0.15 milligram cobalt per pound of body weight daily.

• Adding 1 ounce of raw liver per pound of food may be beneficial.

Table 2
Requirements for a Commercial Food Diet (Moist)
(Increased fat and vitamins; magnesium restricted)

	Percent
Moisture	70.60
Protein	13.00
Fat	8.70
Carbohydrates	5.00
Fiber	9.70
Ash	2.00
Calcium	0.330
Phosphorus	0.230
Magnesium	0.023
Sodium	0.170
Calories	690 per can

---

2. If your cat is **constipated** (symptoms: crying, cessation of defecation, pain when palpated), try this low-fat, high-fiber reducing diet:

Home-Cooked Meal

　　1¼ pounds ground, cooked liver
　　1 cup cooked rice
　　1 teaspoon vegetable oil

1 teaspoon calcium carbonate
Balanced supplement which fulfills the feline MDR for all vitamin and trace minerals.

Combine all ingredients.

Yield: 1¾ pounds
Note: Feed at least twice daily to stimulate postprandial peristalsis. Restrict access to bones, feathers, skin, or foreign material. Provide free access to fresh water. Maintain clean litter box. Eliminate hair impaction in cats by periodic administration of nonmedicated plain petroleum jelly. It's also a good idea to provide an increase in exercise (see "Cat Workouts," page 115).

Table 3
Requirements for a Commercial Food Diet (Moist)

	Percent
Moisture	76.60
Protein	8.10
Fat	2.00
Carbohydrates	6.80
Fiber	5.20
Ash	1.30
Calcium	0.22
Phosphorus	0.11
Magnesium	0.01
Sodium	0.14
Calories	0.8% 360/can

3. If your cat has a tendency toward **blockage**— feline urologic syndrome (symptoms: crying, frequent incidents of crystal formation), consider the following:

Aim
To try to prevent the urinary concentration of minerals found in urinary calculi and crystals and to feed a diet promoting an acid urine pH.

For Prevention

• Feed a diet providing less than 20 milligrams of magnesium/100 calories or containing less than 0.1 percent magnesium in the dry matter. Most commercial cat food labels do not contain a magnesium guarantee. Generally, if the ash maximum is greater than 5 percent in the dry matter of a dry food or of a soft-moist food, the magnesium content exceeds 0.1 percent. There is no relationship between ash and magnesium content of canned cat foods.

The idea is to restrict magnesium, which builds up solid material in the cat's urine, and to build up an acid urine in which the solid material (crystals) are less likely to form.

Some cat foods claiming a "low ash content" contain an inverse calcium : phosphorus ratio, a serious dietary defect. Many cat food labels do not contain a low-ash guarantee. These foods are not recommended for feeding cats with FUS (feline urologic syndrome, or blocked cat syndrome) unless their magnesium content is known to meet the recommendation.

• Feed a diet that results in the maintenance of an acid urine pH.

• Provide clean water at all times.

• Try the anemic cat's diet (page 27), but add B complex with iron (amount prescribed by doctor).

If you decide to go with a commercially prepared cat food, Table 4 shows the requirements that should be inherent in the food.

4. If you have an **orphaned kitten** (symptoms: crying, nervousness), you must replace the milk and care that

Table 4
Requirements for a Commercial Food Diet

	Percent
Moisture	71.00
Protein	12.50
Fat	8.50
Carbohydrates	5.70
Fiber	0.40
Ash	1.40
Calcium	0.20
Phosphorus	0.19
Sodium	0.15
Magnesium (15 milligrams per 100 kilograms calcium)	0.02
Calories	680 per can

would normally be provided by the lactating female. Replacement milk for an orphaned kitten may be administered by a doll bottle and nipple. Try this diet: Feed with KMR (put out by Borden and available in pet stores). Begin weaning to solid food at 3 weeks and mix replacement milk (Borden's KMR diet) with commercial food that has high-quality protein, restricted fat and fiber and easily digestible carbohydrates. Provide adequate vitamins to support growth (avoid oversupplementation). Convert to commercial food at 5 to 6 weeks of age.

5. Here are some more recipes for home-cooked meals. Check with your doctor to see if they're appropriate for *your* cat.

Home-Cooked Meal for a Hypoallergenic Cat
    4 ounces cooked lamb
    1 cup cooked rice
    1 teaspoon vegetable oil

1 teaspoon dicalcium phosphate
Balanced supplement which fulfills the minimum daily requirement for all vitamins and trace minerals.

Discard excess fat from lamb and combine ingredients, mixing well. Do not season.

Yield: ¾ pound

Restricted Protein/Phosphorus Diet for Liver and Renal Ailments

1 pound cooked liver
1 large hard-boiled egg
2 cups cooked rice
1 tablespoon fat (bacon grease or vegetable oil)
1 tablespoon calcium carbonate
Balanced supplement which fulfills the feline minimum daily requirement for all vitamins and trace minerals.

Braise meat, retaining fat. Dice or grind liver and egg. Combine ingredients and mix well. If too dry, add water (not milk).

Yield: 1¼ pounds

Okay, you've determined your cat is in good health and her behavior problem is coming from whim rather than ill health. What next? Simple. Next, we work on training the antisocial whims out of your cat's life to make her the sweetest pet in town.

3

Training Your Cat to Stop Doing Something . . .

The Big Five Problem Personalities

C'mon, Clarence, cut it out! There is no question that it is easier to train your cat to *do* something rather than to *stop* doing something. Let us then tackle the hardest areas of training first—getting your cat to drop the behavior that makes him a pain in the neck.

Although it's true that cats are as varied in personality as people, when it comes to behavioral problems it seems to me that there are only five big ones. Your own pet may overlap and have a combination of two or even three problems (requiring the patience of a virtual saint to convince him he'd be happier dropping his errant ways). If this is so, I suggest you concentrate on the areas one at a time to make training easier for you and more comprehensible to the cat.

Problem Personality One: The Killer Cat

What gets into him? One moment he's the soul of reason and tractability, purring, rubbing, making love to everyone—the next moment he turns into a wildcat, biting the hand that holds his dinner—let alone the guest who makes the mistake of reaching out to pet him one too many times.

Kitten Diplomacy

Let's backtrack a moment: before your Killer Cat grew up, he was a kitten, warm, purring, malleable and eager to learn. Let's just look at him in this stage—between three weeks and three months. He's at his most impressionable, his "imprint" months. He thinks of you as his own parent; you become for him, in modern psychological terms, his "significant other"—that is, the person who protects, loves, teaches, feeds him. You're his own Big Cat, his mentor, and pal. And although he never forgets he's a cat, what you say and how you act now pretty much prescribe his future behavior. *Killer Cat behavior comes about because—when he was a kitten—*

• You never said the "no" word, that lethal, awful "no" word. "No" to the cute, kitteny lashing-out with claws. "No" to the raking of claws on the couch. "No" to the antisocial behavior you'd surely say no to if your cat were a kid.

• You never made positive, "yes" sounds. "Yes" can be any combination of soothing, loving, approving words in the language that you and your cat develop ... it shouldn't be the short, staccato, never-changing sound of the "no," but a slow, low, murmuring kind of yes. "Yes" can be, "Oh, you brilliant, marvelous kitty, you've started clawing on your post instead of the suede chair ... oh, *gooooood* you." As any diplomat understands, the bad guy has to know you mean business. That implies consistency. Kitten diplomacy relies on *never* allowing unsatisfactory behavior to occur without the dreadful "no." If you allow it sometimes, when the kitten's particularly adorable, and don't allow it when you haven't the patience to play with a clawing kitten—you're going to have a crazy, mixed-up cat on your hands—who might just grow up to be a Killer Cat.

• You never understood the value of the bribe. The bribe, in feline language, is always the treat, always something to eat. Child psychologists may tell you from now until next May Day that bribes are not terrific, but when dealing with cats, take it from me, food bribes are the most terrific thing you have going. Your cat is always doing her feline thing, hunting and stalking—always on the hunt, so to speak. What is a bribe to a kid is a very strong training inducement to a cat, so never forget the extraordinary argument that a food reward brings: it's positively eloquent. When your kitten acts aggressively, the "no" word sets the stage. When the claws retract and the biting stops, a gift of food that instant *sets the memory* of what's expected. That's why it's a good idea to have a pocketful of treats readily available when you're playing with (and training) a kitten. If you have to walk to the kitchen and say, in essence, "Oh, here, Tabitha, is a reward for what you did back there ten minutes ago," forget it. So, keep that phrase in mind—*set the memory*. You can set a positive memory as easily as you set a negative one. Your kitten/cat *does* remember positive things if you respond instantly: I guarantee it. Feed your kitten the treat from your hands, by the way . . . it counts much more than if it were dumped in a dish.

• You never "shnoogled" with your kitten. Look, I know we promised you literacy in the first chapter, but shnoogling is a real thing—a mixture of cuddling, snuggling, handling, nuzzling. Cats that haven't been shnoogled with as kittens are basically antipeople. They don't know how good we can be, how comforting we can feel. A kitten that hasn't been shnoogled can't be expected not to claw, bite or fight when she's frightened or worried. Being ignored, she's had to fight for attention, and that could well include very antisocial acts. Not "loving up" a kitten in the season of her greatest learning is a very big mistake. Like a child, a cat has to have a love affair with the human race. Touching is the best expression of that love.

Touching, by the way, is as important in the care of your cat as feeding it. Every cat owner ought to get used to "hands-on." It helps diagnose medical problems at home like abcesses, scabs, cuts, growths and other problems that may cause behavioral reactions as well as physical ones. You can't *know* your cat if you haven't handled it. This is particularly important when the cat has an illness that *must* be treated by handling. One diabetic cat patient was lucky to have had her diabetes diagnosed and under control, temporarily; she was unlucky, though, because her owner couldn't bear to give her the daily shots of insulin she needed to live. Not only was the owner unfamiliar with the cat's "topography" and wary about handling the cat to give the shots, the cat herself, unused to the feel of human hands, could not adjust to being restrained every day for even a minute. Medicine, in this case, could have saved an animal's life—if her owner had been a "shnoogler."

Handling

Okay, let's assume you got a 3 on a scale of 1–5 in Kitten Diplomacy. It's over, there's nothing you can do about it, you wish you'd been smarter then. In the meantime, you have adult Tabitha who needs to be whipped into shape somehow, *right now.*

Whipping into Shape

Don't do it! If there is any act that is anathema to cats, it's hitting, slapping, kicking them. Cats *really* don't like that at all. In fact, it causes them to be pretty hostile.

Although common sense tells us that a cat needs affection and positive reinforcement, much as children do, it's a big mistake to anthropomorphize our cats and expect them to react to a slap on the flanks as a child would. The cat may know it's doing something wrong but doesn't know you still love it even if you give it a whack. It will learn nothing from the whack but distrust of humans. Many of the behavioral patterns of a cat are inbred and instinctual, and punishing it physically only serves to make it respond instinctively with more of the same that prompted the whack. Dogs are different— sometimes they do learn from physical restraint or hitting (although most dog trainers would frown heartily on the hitting part). But cats *never* respond to being hit—their response is flight or counterattack.

(There is also undocumented evidence that cats have an invisible defense mechanism that seems to create an invisible shield against physical punishment. A friend of mine tried to backhand her naughty Siamese but despite accurate aim, completely missed him and broke her wrist against the doorjamb behind him.)

Roughhousing

Don't do that either. Encouraging a cat to act "cutely" wanton is asking for trouble. The moment your kitten or cat starts playing by grabbing your fingers with un-sheathed claws, *stop* playing with it. Say the "no" word. *Set the memory.* When the cat calms down and acts docile once again, speak to it in loving, soft tones; give it a treat. *Set the memory* of Docile. Depend on it: like night follows day, a Killer Cat will always evolve from rough-house play. How is it to know, after all, when wild, aggressive behavior is cute and when it is forbidden? How's that cat to know that his boss doesn't love being leaped on with claws raking over his new shantung suit, when he allowed leap-on in the play session yesterday—in fact, laughed encouragingly at it?

What Can You Do to Train the Killer Cat?

First, understand something about feline aggression: before you train something *out* of an animal, identify why it's there in the first place. That's the only intelligent approach to civilizing your cat. If you had a seven-year-old child who was eating poorly, biting her brother, drawing black circles all over her room—you would attempt to get at the root of the problem before you punished her for ruining the wallpaper. Was she "acting out" because she was jealous of her brother, because the teacher at school yelled all day, because the kid next to her stole her snack three times a week? You'd have to find out before you could handle her antisocial behavior, because if you just assumed she was instinctively destructive and took away her television privileges, you'd accomplish nothing—maybe exacerbate the problem. Ditto with the cat.

There are reasons why cats become belligerent Killer Cats. Here are the reasons—and the remedies.

Fear-induced Belligerence

A cat that feels restrained, confined or frightened will almost certainly attempt to attack. A cat that feels threatened by an approaching individual (the cat doesn't *know* how terrific your mother is) will lash out. He will feel the need to be defensive. This is probably the most common reason for cat attack. You will know it as the underlying cause if you carefully observe the physical posture of the cat as he attacks:

- Are his ears drawn back and flat to the head?
- Is he crouching with his head drawn in (instinctive behavior to protect the nape of the neck)?
- Does he spit or hiss?
- Is he arching his back and turned sideways?
- Is his tail waving back and forth?
- Has he rolled over just before he strikes with his forepaw or teeth?
- Are the cat's eye pupils dilated?

If any or all of these physical statements are made by the cat, assume that he is aggressive because he's scared silly. Don't read any of these signs as submissive behavior: they're not.

Remedy: Brainwashing. A rather intensive, two-to-three-week program of brainwashing does much to eliminate hostile behavior in a healthy kitten. Brainwashing consists of repetition (as you remember from the stories of the returning prisoners of war). Repeating the same action over and over again until capitulation occurs is the theory behind brainwashing. You're not attempting negative but positive reinforcement, so the hostile, scary elements you'd think were part of the process are not. Only the positive, encouraging effects remain. If the cat is hostile because he's shy, every day, as many times as possible, ask strangers (to the cat) to approach the cat gently, softly, lovingly. You may have to bribe your mother to do this, but the results are well worth the expenditure. The "stranger" approaches the cat *very slowly,* hand out in an offering of friendship for the cat to smell, inspect. At first, the "stranger" doesn't touch the cat—just approaches it and stops about a foot away. Gradually, slowly, the cat is touched, then stroked, all the while being spoken to soothingly, lovingly. If the cat strikes out, the word "no" is firmly said. As the "stranger" says no, she should (or you should) "cup" the cat's face with one hand firmly placed over the frontal bone and top of the mouth, and the other hand, under the chin. This reinforces the no. If the cat relaxes and seems temporarily pliant, a treat is immediately offered by the hand that "cupped" the cat's face.

Never restrain or hit the cat or even tap it aggressively: that won't work—it *never* works, and may only exacerbate the problem when the cat is shy or hostile because of fear.

Gradually, Killer Cats who attack from fear begin to trust. Your soothing voice provides an equivalent of hands-on comfort—before you can get the actual hands on. Positive vibrations—that you care, that you're not going to hustle it—begin to get through to the cat's psyche. She begins to trust approaches, people's hands, strangers. She associates a treat with allowing herself to be touched and approached. If the cat is young, brainwashing takes a shorter time. Older cats require more patience, more time, a willingness on the part of your recruits to suffer some scratches and bites in the interests of science (that's how you put it to the volunteers), and they, too, can be trained. Only a very few cats are incorrigibly "killer" and then I've found limited and sporadic success using chemical tranquilizers during the training program. If worst comes to worst, discuss this approach with your own veterinarian but never attempt to administer tranquilizers without medical supervision.

Remedy: Spitting Back. A cat's pawing, biting, hissing and spitting can be counteracted by a little of its own language. *You* can spit back. Now, if you do it with your mouth, the cat will think you're nuts, and you would be; also, it will be able to identify you as a foe. The way to do it is with one of those cheap plant atomizers or water guns that you buy in any five and dime. One blast with a water gun tells the cat a very emphatic no. (A little cider vinegar added to the water *really* repels the cat.) If the behavior is repeated, another blast reinforces the lesson. The good part about all this is that the cat doesn't figure that you, its beloved master, is spitting at it . . . the water gun gets the blame. It's a very effective argument. Water gun spitting will be used in other training procedures. Try it with the Killer Cat.

Remedy: Hissing Back. An enormously effective alternative to the water gun or atomizer is the *air gun*. It's particularly useful when you don't want to soak the furniture, your guest or the cat herself. Cannisters of air can be purchased at photographic supply stores (they're used for cleaning cameras) and they generally have nozzles which can be adjusted for intensity of spray. Cats *hate* that whoosh of air directed at them because it so resembles an angry hiss both in sound and feeling. An air gun can be used in discipline wherever a water gun is warranted: it's less messy and somewhat less forceful, although it certainly gets a message across when employed in conjunction with your firm "no." (Water or air guns can be used on "Killer Cats" who respond aggressively for any reason.)

Remedy: Dig In There. Cats sometimes act very belligerently when they're frightened that they've lost control—just as people do. Sometimes, a skittish cat who's placed on a smooth surface for grooming or for medicine administration will claw and flail about in fear. If you have a cat who responds violently to being restrained, try putting a small piece of rug (even a bath mat) under it so it can dig its claws in. Somehow, that gives it a feeling of security. Or, wrap its whole little body in a warmed towel to ease its fears. Again, I emphasize love, touch, gentleness, caring, talking as the finest approach.

Remedy: The Cat Expletive. During any behavior training technique, you've taught your cat the meaning of "no," of course (used only when the cat's done something wrong *in terms of behavior.* Never use "no" when teaching tricks or training. Positive reinforcement when the cat's done something right is the way to go when training). There is one more sound that is hugely effective, if not overused, and it's the equivalent of the human four-letter-word expletive. It's *psssst!* The noise of this expletive is as close to the sound of a cat spitting that a human can muster. If you use it only in the direct circumstances— say, as your cat is leaping toward your neighbor's visiting baby—most cats will respond with immediate compliance.

Territorial Belligerence
This is the second reason why your cat may be a Muhammad Ali—fighting you, your guests and perhaps all other cats and dogs, besides. The physical statement that says, "I'm protecting my space, fella, and you better get out of here, pronto" is this:
- an aggressive bounding right at the "enemy" and a round of "boxing" (paw feints and blows);
- the cat crouching and staring you down;

• the cat, if another animal or person becomes too aggressive, actually assuming a Muhammad Ali stance, standing on its hind legs and pawing or boxing with whomever it suspects is threatening.

A little background information on territorial claims: traditionally, through the ages, cats have marked areas that they decide belong to them. Not all cats become territorially belligerent, but most will. What complicates matters is that a cat may live in perfect peace for a year or two, then all of a sudden develop a "this is mine/that is yours" complex. Territorial belligerence can occur when a new cat is introduced into the home or when several cats, raised happily together, reach maturity. *Or,* territorial belligerence may come about as a result of:

The Barnyard Effect. I know that sounds like a Ludlum mystery, but it's really quite simple and quite fascinating. Between the years of 1977 and 1981, a Cornell University study team monitored the daily doings of cats on four farms. Among many things they discovered, one important finding was that too many cats, "a dense population" in the barnyard, too much crowding, causes cats to be much more belligerent in claiming their space rights. They hiss, they scratch, they bite, they fight. In this society, with very little privacy, cats feel as if they've *got* to fight to keep a bit of space inviolate, and their own. It's really quite understandable and a very good argument against a city dweller in a small apartment overpopulating the place with pet cats. The Barnyard Effect produces a very angry, hostile feline.

Remedy: Respect the Boundaries. Simple as that. Guests should be told not to sit on the one chair in the house that the cat seems to have marked as his own—or go to

the cat when he's in his own corner. You know very well
which corner, which chair is your cat's special place:
he's almost always in it—and in times of stress, he *is*
always in it.

   Remedy: Blur the Boundaries. This takes time, but it's
a wonderful remedy. My clients who have Muhammad
Ali Killer Cats have reported the greatest success from
spending at least a month or so working with this
remedy. The goal is never to let two cats (or a kid and a
cat) fight it out for dominance. That is useless. No one
ever wins. The fight just goes on. But, you can blur the
boundaries at an infinitesimally slow pace so that the cat
doesn't realize he's being coerced into sharing his territory. Blurring the boundaries is necessary when a cat
(or *cats*) seems to have usurped an unfair or inconvenient
amount of space and is intent upon preserving all of it.
   Every day, intrude just an inch or so into the area
of the cat's domain. Don't go up to the cat—just get your
scent (or a guest's scent) into the area. If you have two
or more cats who are hostile to each other, separate
them in different rooms. First, take the intruding cat and
place him in the other's territory to walk around and
leave his scent. Do the same with the second cat.
   Gradually introduce visual and auditory sharing: bring
the cats, each in his *own* box, into the areas in question.
Let them get used to the smell and the sight of each
other in their territories. Then, with a barrier erected in
between, let each cat out of his box to patrol the area.
Gradually decrease the size of the barrier, lowering a
wall, or opening a door wider. Take turns allowing each
cat in the other's living area until they can tolerate each
other. And always accompany each training session with

soothing, encouraging talk to each cat, using the cat's private name many times.

Remedy: Share the Boundaries. This applies to bringing an entirely new cat into the home. This technique will help avoid territorial belligerence from happening in the first place.

1. Start with understanding that the "established" cat is the one who's going to need the bulk of the attention. The best thing you can do when introducing a new cat is to select a kitten, or at the very least, a cat of the opposite sex. Remember that this is a potentially stressful situation and the established cat will need, for a long time, to be the "boss," to get the most attention, to get the sweetest talking to.

2. Have someone other than the person who lives in the house bring the new cat in. If you bring in another, "intruder" cat, expect a whole lot of hissing, fighting and biting to express the established cat's jealousy. Choose one of the following ways to bring in the cat.

• Keep the cats in adjoining but separate rooms for a couple of days. Let them sniff each other through the door. Then slowly, with a neighbor carrying the new cat in, have a direct introduction with supervision.

• Or, put the new cat in a carrier and let the house cat sniff him for a few days.

3. When the cats finally meet, make sure you employ the "no" word immediately if there is any friction on either cat's part. Let them know from the start that fighting is unacceptable. If "No" doesn't work, employ the old water gun at displays of hostility. Stroke and talk to the house cat constantly, making it feel loved and secure. Say "No" to the younger cat if it invades the

house cat's private place—that helps your house cat retain its authority. During the lives of both cats, one, usually the first, will always retain some dominant authority over the other. That's the law of the jungle, you might explain to the newer cat. They may eventually share a communal water bowl or a communal litter box, but their separate areas (particularly the boss cat's) will always remain inviolate. Think of it this way: the happiest husbands and wives have one place that's usually their own space, a place to escape, exercise, dream or relax in. Why shouldn't cats have the same?

4. Keep this in mind: don't over-react to what looks like fighting but might well be roughhouse play. Cats need to form their own relationships and if you keep breaking up their play because you're afraid they'll harm each other, you won't allow them to form the master-disciple relationship many cats enjoy. It's not unnatural for one cat to submit more than the other, and, also, it sometimes takes time to come into one's own. I once had a client who introduced Shayna, an apricot shorthair, to Kafka, an Abyssinian. Kafka was a bully, a brute, and really drove Shayna bananas. One day, Shayna had simply had it. She moved out of the house, into the garage, and wouldn't set foot back in the house for two years. If she was brought in, she'd escape through the cat door as soon as possible. Two years later, she came back. But, boy, things were going to be different around here, she let everyone know. She was on the attack this time, and when Kafka got obnoxious, Shayna responded with a fury. She never took second-class citizenship again, from anyone. As a matter of fact, my client reported, every time she yelled at Kafka for one reason or another, Shayna would hear my client *from three miles away,* and

hissing and biting she'd come to help in the chastising. The two had an uneasy truce from then on.

5. Finally, one of my clients has had enormous success with *lightly* spraying a new cat with cologne and then spraying the established cat with the same cologne—worth a try, I should think.

Intermale Belligerence

This is the kind of anger and hostility males have traditionally carried for other males who infringe upon their lives. Wars start from intermale belligerence. In the case of cats, wars are probably avoidable. You will recognize this kind of Killer Cat activity by the confrontation behavior of the two cats. It's awesome. The aggressor (or newcomer) will advance slowly, turning his head from side to side. Sometimes, he'll stand up on end on his hind legs. The cats stare at each other. The ears are not flat, but often raised. After a long face-off, one may bite the nape of the other's neck and the battle begins. You've got to do something.

Remedy: Castration. Do it. Sounds harsh, but male cats who are pet cats must be castrated, in my opinion. This makes for an infinitely nicer cat, better smelling, pleasanter, more sociable. It is not an anticat act to castrate it. It does not violate your cat's essential catness. People who have lost their reproductive organs are still people. Research shows that approximately 80 to 90 percent of fighting cats turn out to be great pals after castration.

Male gonads *compel* cats to fight. You're not only doing your cat a favor when you castrate him, because he'll find living with people so much easier, you are also cutting down on a vast number of unwanted pets who wander the world hungry, sad and constantly in danger. It is possible, of course, that castrating (or neutering) your male cat will not completely alleviate his tendency to pick a fight, but it should help enormously. Castrate him young—about six months of age—before spraying and fighting become a habit. In severe cases where *nothing* works to alleviate intermale belligerence, and you want to keep two male cats, your veterinarian can try hormones or tranquilizers, but that's rarely necessary after castration.

A word about female spaying, while we're on the subject. It's also an excellent way to make a female a more pleasant house companion—and to cut down on unwanted litters, for which, I might add, you ought to consider yourself totally responsible—should your cat remain unspayed. Responsible means keeping them or finding good homes for them . . . not destroying them or letting them loose somewhere! Contrary to myth, neutered (the term applies to both spayed females and castrated males) cats do not become fat or lazy.

Spaying a female cat should take place before her first heat or immediately afterward. She does *not* have to go through one heat to be a satisfied animal, contrary to rumor. Spaying is a routine and safe operation which takes about forty minutes and promises a much more well-adjusted house cat. Incidentally, spaying often discourages the growth of mammary gland tumors—a common cat ailment. Veterinarians have discovered that cat breast cancer is found more often in unspayed cats because the estrogen hormone receptors in unspayed cats encourage tumor growth. (This may be a reason why the hormones that are present in some birth control pills for women have been held by some scientists to be accountable for breast tumors in women who take the pills for long periods of time.) An infected uterus will be seen far less often in those female cats which are spayed than in those that are not.

So, the best remedy for intermale belligerence (and female belligerence, also) is neutering your cat.

Remedy: Chemical Soothers. Your veterinarian, at last ditch, can try the use of a psychoactive, tranquilizing drug such as Valium to calm cats who are aggressive for territorial or intermale reasons. One of my clients married a man with a cat whom the client's cat despised at first sight. Nothing helped, not separating them and trying socialization slowly—not respecting boundaries—*nothing*. Neither client would give away the cat he/she came with. It was the marriage or the cats. I had my work cut out for me. I advised them to mix crushed Valium tablets (one tablet a day for two and a half weeks; after that, gradually reducing the dosage) into both cats' food. Your own doctor will calculate the dosage according to your cat's weight. My clients were then told to keep one cat in a carrier and allow the other out to sniff and explore around it—which it did. Hissing and growling at the carrier door took place for a while, the clients reported, and gradually that disappeared. The big day came: both cats were allowed free in the same room. They sniffed each other and then virtually ignored the other's presence, except at feeding time when each had to be fed from a different dish in a different room to avoid recurrences of the angry behavior. After a while, even that changed, although each cat always required her own dish even if they were side by side in the same room. They never became best friends, but learned to tolerate the other. I don't recommend drug therapy as a general rule,

but when the stakes are high and the chips are down—
it's worth a try.

Sometimes administration of a progesterone-like female hormone controls spraying and aggressiveness, and
a male hormone may bolster the self-confidence of a shy
cat underdog (if you'll pardon the expression).

Play Belligerence

Sometimes aggressive and unpleasant behavior can be
seen in cats who are sweet as can be and only out for a
good time. When cats play, they utilize many instinctual
predatory movements like stalking, pouncing on and
biting things that move attractively. It's a wonderful
experience to watch a kitten stalking a catnip mouse, and
many owners, including myself, spend pleasurable hours
indulging in horseplay with their kittens and cats. The
trouble begins when the cat is suddenly ignored, because
its owner has discovered she has nineteen clients waiting
for wise words. Result: play deprivation. Play deprivation
irritates cats more than they can say (or, rather, show,
since we know cats do not talk). They very often respond
to being ignored for long periods by nipping the heels
of a guest, clawing the moving target of a baby, and
stalking the teenager's boyfriend. They become frustrated
and belligerent and attempt to get their play pals back
by attacking them—in a spirit of conviviality, sure, but
attack, nevertheless.

Remedy: Instant Retaliation. Either a little water or
air gun therapy (a splash of water or air) on the nose
does wonders to teach that playmates don't like being
attacked. Or try a brisk tap on the nose (just like the
mother cat used to do to punish errant pusses). Both of
these actions must be accompanied by a resounding

"No" to set the memory. I can't overstress the importance of that no for every possible kind of objectionable behavior. Consider also, as a more humane remedy, spending prescribed amounts of time with your cat daily. Like babies, cats who are expected to be sociable must have socializing.

If you expect a loving, pliable cat, you must treat it as such because self-fulfilling prophecies abound in cat-rearing. Certainly an environment that includes play toys like rubber mice and balls is stimulating for younger kittens especially, but perhaps the most satisfying play of all is affectionate stroking and caressing. Touch is the primal source of affection. The mother cat touches her cats, licking, grooming, caressing them right from birth— an act that not only stimulates their nervous and circulatory systems, but encourages their complacency. If your cat is belligerent and you suspect it's because she feels play-frustrated, cut down on the wildness by starting to talk low and lovingly.

Remedy: Shake and Face Hold. I've told you never to hit your aggressive cat. But, that doesn't preclude *shaking.* When a baby cat is recalcitrant, a mother cat will seize it by the scruff of its neck and shake it. It's a good idea to wear gloves as you do this, and never *lift* a full-grown cat in this way—the neck cannot easily support the cat's entire weight and you could break its neck. But grasping it by the scruff immobilizes it temporarily, and a not-too-vigorous shake tells the cat you're peeved at its play behavior. With your gloved hand, you might also cover its face with your hand—*firmly.* This is a way of disciplining that the cat understands; when you've held it thus for a second or two, release it. In any training for behavior aggressiveness, the idea is to respond to poor behavior *immediately:* never let it go for even five minutes. The cat will never remember that the shaking is for its lashing out at your mother (or defecating on your guest's pillow) five minutes ago. To a cat, five minutes ago is Plutarchian history.

Remedy: The "What's In It for Me?" Tack. Why should the cat stop being belligerent in his play if there's nothing good to come of his newly calmed behavior? When you are playing quietly with the cat, and he's gentle and sweet and just as you'd like him—tell him so. Coo to him, give him a treat. He'll soon make the connection between quiet and reward, and will see that purring equals treat and lashing-out-wildly equals "No!"

Remedy: Consistency. It's important that your cat understands that something is bad all the time or good all the time. If you allow Bumbler to jump on your lap and knead away when you are wearing gardening jeans, but yell at him when he does the same thing when you're in your best designer jeans, he's going to get very confused. How is Bumbler to understand which are the "Calvin Kleins" and which are the gardening jeans? Can he jump on you when he's dry and clean? Then, how come he can't when he's just come in from a muddy terrain? Either you allow your cat to do a thing—or you don't. There are no fine points or extenuating circumstances.

Remedy: For Kittens. Kittens and young cats often bite because they are teething—no more sinister reason than that. Give your kitten something to chew on to stimulate the gums and make those teeth come in faster. I suggest some beef jerky or some raw beef bones (but *never* cooked bones, which splinter easily). Chicken or rabbit bones are always a no-no—raw or cooked.

Finally, for problems that will *not* go away, again I urge you to speak with your veterinarian about medical treatment with tranquilizers or hormones. It's a last-ditch resort before you give the cat away.

Problem Personality Two: The Uncouth Cat

Do you have the only cat in America that refuses to use its litter box, sprays all over the place, knocks things off tables and in general messes up the joint?

First of all, remember that it is against the essential cat psyche to be untidy. When I worked with the big cats in the zoo, if a leopard or a lion was careless about its living quarters, if it dribbled its food all over the place—and left it there—that would be a sure sign that the big cat was neurotic, sad or ill. It happened, alas, far too often, and today we know that close confinement in cages makes cats very bored and neurotic. The more modern zoos have ranges or roaming areas and, as a result, the cats are infinitely cleaner.

Cats *love* to groom themselves and are fastidious about eliminating in the proper place. One of the nicest things about getting a cat is that the owner doesn't have to break his neck "housebreaking" it; the feline nature is a built-in housebreaker. So, if your cat is soiling indiscriminately, if it seems slothful and careless, and if *you've eliminated medical causes,* well, your cat's piqued at something for sure, or, perhaps he's desperately lonely. Just as a child cries out for attention by smearing his feces over the bedroom wall when he knows his parents will be less than pleased, a cat, when angry, lonely or unhappy, will act similarly—even down to dropping his excrement outside the litter box—the most uncatlike act of all.

What Is the Way of the "Couth" Cat?

Cats in the wild bury, with compulsive thoroughness, their urine and feces—especially if they've evacuated within their own territorial boundaries. A happy house kitten or cat needs no brilliant trainer to teach it how to use a litter box. All you have to do is provide the litter box, show the cat where it is, perhaps place some feces from your other house cat (if you have one, if not, forget it) in the box to make your message loud and clear. Then you put the cat in the box and scratch his front paws in the litter. Do this a few times. That's it. Your cat is trained. Now all you have to do is sit back and wait for it to be meticulous. One or two kitten accidents, maybe— but the kitten is essentially housebroken.

Diagnosing the Problem

But what does one do if a kitten so trained begins to get careless? First, understand the reasons for the careless- ness. Sadness can cause a cat to urinate or spray outside of the box. So can anger. So can desire to mark its territory. (Cats mark in three ways: scratching with their paws, rubbing with their cheeks, and spraying—all seem to leave a scent, which makes you understand what an olfactory creature the cat is.) Fear or boredom or attrac- tion to the scent of another cat can cause a cat to be uncouth and defecate, spray or urinate outside of his box. Here's the way you diagnose your cat's problem— and remedy it.

Sadness at a Perceived Loss

Have any of the following recently happened in your home?

- The death of someone in the family
- The long absence of someone in the family
- The birth of a baby
- The acquisition of another cat (or dog or person)

I once had a patient whose owner got married—without asking the cat's opinion on the matter. The new husband set about making himself comfortable but soon became very uncomfortable as he stepped on cat feces every morning. His wife was baffled—this was her trusty childhood cat. He'd *never* had an accident! Certainly, he'd never knocked anything off the table before—grace was his middle name—but objects (particularly the new husband's possessions) were mysteriously shoved off chairs, tables, bookshelf ledges. The situation became critical. Again, it was either the cat or the marriage.

Remedy: TLC. Both were saved. What was needed was an overdose of TLC (tender, loving care), particularly from (in the cat's view) the intruder. The husband, who was a nice enough sort after all, spent many hours playing with the cat, brushing its long hair, scratching its ears. When he'd arrive home from the office, his first greeting would go not to the wife, but the cat . . . "Hello, you gorgeous creature, you're looking so sexy, you're my baby, all right" . . . and a whole lot of other affectionate nonsense. It was he that opened the can of cat food, he that gave the treats and it was the wife who was delegated to capture the cat for the visit to the veterinarian. Gradually (it took about three months, I was told), that cat responded. No more "accidents" of any sort. Cats grieve at the real or emotional loss of a loved person—and they may easily react by doing something contra-cat.

Anger or Pique or Just Plain Nastiness

Sometimes, a cat breaks his litter box habits because he's angry rather than grief-stricken. He expresses his fury by many means from refusing to eat to being messy about his personal habits. I once had a patient who delighted in urinating *on,* right *on,* the playmate of his special child because he was jealous of her. This particular playmate was terrified of him, which made matters worse. Whenever she entered the cat's domain—whooosh, she'd be bathed. It was terrible. Instead of reacting immediately to the cat and chastising it, the cat's owners became involved with apologies and cleaning up the mess, and the cat was home free.

Remedy: Firmness and Instant Action. It wasn't until the owners became ready with trusty plant-atomizer spray action and a firm "No!" that the cat began to refrain from being obnoxious. I explained to my patient's owners that a cat will often take a dislike to a human whom they sense fears it, and when this happens, swift retribution for the cat's overt hostility is required. Cats often "mark" frightened people by rubbing against them to leave a scent. Try a firm "No!" for this cat also, if it's unpleasant to the recipient.

Urine Marking

Although it occurs more frequently in uncastrated and unspayed cats, spraying (or marking of territory by urinating on it) can also occur in neutered cats. Surely, if you have an unneutered animal and he or she is urinating outside the box, get him or her neutered immediately: it's a kind and intelligent act (unless you really want kittens first).

Remedies: Solving the problem of neutered cats' spraying starts with figuring out the location of the spraying. If shoes or clothes of one member of the family are constantly being sprayed, the cat may be fearful or angry at that person and a truce must be made. Cats will start just "missing" a litter box, or spraying outside a box because of a social stress problem—perhaps too many cats are using it, in which case you must provide separate boxes. Cats also spray outside the box if the box itself is not clean enough.

I have had clients say they've met success encouraging their cats to stop spraying by using two other methods I suggest. One is to hang a sheet of aluminum foil by fishing wire or rope in front of the object that the cat likes best to spray; somehow, the noise of the spray hitting the foil turns some cats off and they soon stop spraying. Another method is to feed the cat right at the spot he's in the habit of spraying. Cats do not love eating where they urinate, and this often works to curb the spraying. **Caution:** any time you use rope or wire, make sure the cat won't eat it or get tangled in it!

Because so many factors, including hormonal, environmental and behavioral variations, influence spraying for marking, you really should include your doctor's advice in your training methods—sometimes progestins temporarily administered can reduce the problem if your cat continues to "mark" after being neutered, and there don't seem to be any problems you can pinpoint.

When There's No Apparent Cause

Sometimes, it's ours not to reason why. Even if you're the intelligent cat owner, you just can't try to creep into the cat's head *every* time and figure out causal possibilities. Sometimes, you just have to try to make it better without too much introspection. Whether your cats urinate, defecate or do both outside of the litter box is academic. The important thing is to stop the behavior by correcting it *immediately,* and to be, above all things, consistent. Your cat must know that it is intolerable for it to act in this manner and that it will *hate* the repercussions if it continues to do so. First, let me give you some handy memory enforcers: they're not punishments but they serve to remind the cat that what he's doing, or about to do, is not terrific. Here, then, are some behavior modification remedies, to wean the cat away from the wrong area and into the right.

Memory Enforcers

If your cat has urinated or defecated in an area outside of the litter box, it will tend to return to the same spot because of the smell. *Make that spot awful for it:*

• Spread double-faced Scotch tape around the area: the cat will be very unhappy about returning to a place to which it sticks.

• Soak a cloth with white distilled vinegar and put it in the area. Carry the cat to the spot and put its nose close to the cloth. (Don't ever rub its nose on the cloth or *call* the cat to come. Never call a cat for an unpleasant purpose—it will never again come when called.) The cat will associate a strong, unpleasant smell with soiling the wrong area.

• Place aluminum foil, wax paper or a plastic sheet across the area that the cat defaces. Cats don't love it when their urine's not absorbed, as it is in litter or sand or on a nice absorbent rug or bed. They also don't like to have damp feet.

And make its own box infinitely desirable by keeping it clean, providing a pleasant substratum (perhaps soil or sand if your cat was an outside cat), or providing separate litter boxes for more than one cat.

Behavior Modification Remedies

Noise Therapy. Naturally, every time your cat soils in a wrong place, and you see it, you'll respond with a loud "No"—and *noise therapy*—loud noises are particularly unpleasant to cats. The loud swat of a rolled-up newspaper on a hard surface, the shriek of a whistle, the clap of a hand when you catch the cat in the act tends to get the idea across. Always, then, after the noise therapy, pick the cat up and return it to the litter box as if to say, here *not* there.

The Substrate Switch. The litter or sand you place in the elimination box is called substrate. When cats eliminate, their general pattern, if you'll watch, is to scratch in the material to make a nice, comfortable little depression, eliminate in that depression, and then cover up the mess with the substrate material. Sometimes, a cat decides it doesn't like the substrate you're providing and will scratch on a hard floor near or even on the sides of the box, thus teaching itself to eliminate on hard, smooth surfaces instead of sandy, littery ones. The trick is to reteach it to use the litter substrate. First, take all the litter out of the box and make it smooth, thus matching the hard surface the cat apparently now prefers. Gradually start adding the litter until the cat begins to associate *litter* with elimination—instead of *smooth* with elimination.

Sometimes the cat will start to defecate or urinate on a bathroom rug or even a living room rug. Try placing a piece of rug right in the litter box for the cat to scratch—and still reinforce the idea that *here* is where your toilet is. Gradually, remove the rug substitute by making it a smaller and smaller piece until the cat transfers its scratching to the regular litter substrate.

What if your cat really hates sand or other commercial litter? It happens. Try alternates—shredded paper, clay, earth, wood chips—until the cat finds one attractive to him. If the new substrate is inconvenient to find or purchase, gradually mix some of the available litter in with the cat's preferred material until the convenient one reigns supreme in the box. If your cat absolutely refuses it, you'll just have to stick with his favorite. Cats need *some* victories.

Change a Painful Association. Your cat may have had a bout of cystitis (a bladder infection) in which she's suffered great pain when using the litter box. She may, thereafter, associate pain with the box and try to avoid pain by avoiding the box—even when her cystitis is cured. In this case, try changing the location of the box as well as the litter material in it: by disguising the toilet, you may entice her back into the bathroom.

Chemical Solution. For confirmed sprayers, I've had quite a bit of success with the use of long-acting progestins (such as Depo-Provera). Progestin can be used in both intact and castrated male cats and spayed females and it has the effect of also reducing aggressive behavior and roaming tendencies. (I would not advise using progestin on an intact female.) As I've said before, chemicals are not to be tried, in my opinion, until all else has failed.

A Little More Attention to My Box, Please. There are four big reasons why cats often spurn their litter boxes:

1. The boxes are too deep. Boxes that are too deep are often a hassle to get into. The cat tumbles headfirst into her litter or, once in, is half buried by her own excrement. Not great, she figures.

2. There's too much litter in the boxes. When there's a copious amount of litter in the box, urine tends to spread all over the litter instead of sinking down to the bottom in one easy clump. The cat thus finds it difficult to cover the wet spot, which is her natural inclination. Every time she goes in her box, her ankles get damp. Not great, she figures.

3. The "privacy" top is a pain in the neck for him (literally). Low, covered tops on boxes cause the cat to hunch over as he squats to eliminate, and this, in turn, causes bladder and bowel impairment. Not great, he figures.

4. The boxes aren't changed often enough. When *you're* careless or messy and don't change the box at least every other day, don't complain if your cat refuses to visit it. He'd rather eliminate on that nice, clean floor. Wouldn't you?

To sum it up: the lowest, simplest litter pan is the best—a shallow, commercial dish-washing pan is best. If your kitten has trouble getting into the box, put a ramp up to it. You should put litter directly into the pan in reasonable quantities—not huge, thick amounts. If you line the pan with newspaper or plastic, that will only make it smell. The litter itself covers and blots out the smell, but the urine just soaks into that newspaper or plastic, sitting there—smelling up the area. Don't cover the box. An uncovered box will allow the cat free movement as well as allow *you* to see what's in there, so you can clean it out frequently (a scoop hanging near the box is handy to lift out new feces and urine clumps and flush them down the toilet—also cutting down on smell).

If you get rid of waste as soon as you spot it, the box can then easily wait a day or so, or even three, if necessary, before it's changed and scrubbed out. Speaking of which: it's a great idea to wash that box with soap and water every time you clean it. This cuts down enormously on bladder infections and odor. A perfumed deodorant spray certainly won't do the trick: that's a little like perfuming your own body when it smells from perspiration. Also, cats avoid chemical odors. In short, if you take care of the litter box responsibly, there should never be an odor in your home—and, your cat will love it!

If He Doesn't Love the Neighborhood, Change the View. Sometimes a cat acts uncouth because he hates the *location* of this box. Many people put the litter boxes where the cat eats or sleeps. *Think:* Would it please you to eliminate right near your bed or table? The cat doesn't think much of the idea, either. Try putting his box in the bathroom—the most appropriate place, anyway, since the toilet's handy in which to scoop out clumps of urine or feces when you're in for your own pit stop.

Sometimes the cat just arbitrarily decides he doesn't like his litter box in the bathroom and chooses the living room instead. Change his mind by gradually making the living room a place of play (keep his toys there), or a place of eating (place his food and water there); the cat will rarely want to eliminate where he plays or eats. You can also try some foul-smelling substance (foul to the cat, that is) like vinegar soaked on a rag in the living room where the cat's been eliminating. Perhaps the cat will then return to his box in the area of *your* choice—the bathroom. Sometimes, as a last resort, you can stoop to trickery. Trickery is great, by the way. Never be so respectful of your cat that you will not stoop to tricking him. If he insists upon eliminating in the living room (or any other inappropriate place), and nothing will change his mind, try placing his litter box right in that area for a week or so until you train the cat back *into* the box. Then, gradually, only an inch or so a day, move that box back into the bathroom until the cat is eliminating *in* the box, *in* the bathroom, hardly even aware he's been tricked. The cat will like the view *you* choose for him, eventually. By the way, don't change the location of the litter box as easily as you change your decor. Everyone likes his bathroom to stay put.

Mask Smells. Cats are prone to eliminate where other cats have had accidents. You must eliminate any feces or urine smell of another cat by scrubbing the area carefully and then marking it with white vinegar to repel the second cat. Try an enzyme odor remover that you can buy at most pet shops to really get that odor out, and don't forget to clean the pad *under* the rug as well as the rug. A last resort to rid an area of a urine smell is to spread a mixture of baking soda and aquarium charcoal over the area—cover up the mixture and allow it to remain for a couple of weeks before you vacuum it up. That should do it if anything will.

Hey, Isn't This Where I Got Hit in the Head? Sometimes cats avoid a litter box because something violent or scary happened to it in that area. One of my clients said that a roll of toilet paper dropped onto her cat when he was squatting in his box and nothing could make that cat go near the bathroom again. Other cats have been punished or yelled at when in the litter box and then associate it with unpleasantness. Some cats have even been given medicine in the area near the litter box and that puts the kabosh on the area. Nothing to do but change the location of the box to a different area. Gradually, after he's again comfortable with the box itself, you can use the "inch-by-inch" method of returning it to the bathroom (or your desired location). It may be that the cat absolutely refuses the old location—forever. So, be a sport. Find a new, permanent place for the box. Know this: when animals are defecating or urinating, they are in their most defenseless positions. Atavistically, it was the time when it was easiest for an enemy to attack. Therefore, it makes sense to put the litter box in an area where there are not many threats, loud noises, or traffic patterns to discourage the cat's consistent use of the area. How would you like to have your toilet in the center of a tennis match?

In the Tub? Oh, No! Simple, keep an inch or so of water in the bathtub to discourage the cat's using it as an oversize litter box.

The Isolation Cell. It's worse than it sounds. Sometimes, a cat simply has to be confined to a small area that holds the box, his food and water and a few toys. Visit him frequently of course for love and attention; eventually, when he gets the idea of the box—and only the box—he can gradually be given greater freedom.

The Great Encompasser. Let's go back to that tub for a moment. If your cat just simply misses her box much of the time, or trails waste material out of the box and it's driving you crazy, try putting her whole litter box right in that dry bathtub: the tub will encompass any spillage and will also provide privacy for sensitive felines. Do I have to tell you to remember to remove the box when you take your bath? I didn't think so.

And, finally,

The Brilliant Solution—Toilet Training. If your cat is having neatness problems (and even if it's not), this might be a great time to consider training it to eliminate in the toilet. Many owners are ecstatic at not having to clean out litter pans and are overjoyed by the *absolute* lack of smell in the house or area. It helps if you have a toilet that's pretty much for the cat's use, alone. It won't do to have your kids and the cat fighting for bathroom privileges and/or it's not so pleasant for a guest, or even for you, to always have to use a toilet filled with cat droppings ... and sometimes even the best of trained cats miss.

The cat's natural squatting posture makes it easy to take proper aim on the toilet and many cats find toilets an absolute snap.

A Word of Caution: Be patient. It takes how many weeks (months?) to toilet train your kid? Don't be impatient if your cat doesn't get the hang of it right away.

Here's how you can toilet train your cat:

1. First, make sure your cat (it's easier to start with a young one) is completely litter box trained.

2. Get a round, shallow, aluminum pie tin that will fit into the opening of a spare toilet seat. Affix it very securely to the seat and place both on the floor. Fill the aluminum tin with litter. Let the cat get used to this new litter box-attached-to-a-toilet-seat on the floor for a week

or so, encouraging him to use it by scratching in the litter with your hands and speaking to him gently.

3. Gradually, over a period of a week or so, raise the toilet-seat-litter-box off the floor by placing it ever higher on very solid objects (a specially constructed wooden platform with legs that expand gradually is best; telephone books secured to the floor are okay in a pinch). The idea is not to allow that litter-box-toilet to fall when the cat jumps on it—one mishap and the cat's off toilet seats forever. By raising it slowly, gradually, you will bring it to the height of the toilet seat and train the cat that it has to jump to use the box.

4. Finally, after the cat has used the box and seat successfully for a couple of days, remove your base and put the aluminum litter pan on the *real* toilet. Let the cat use it for a few days.

5. When the cat is comfortable jumping onto the toilet seat, make a small hole in the aluminum litter tin. The next day, enlarge the hole. Gradually, imperceptibly, make the hole larger and larger until the cat is straddling the toilet and his feet are firmly planted on the seat. (*Note:* Always leave the seat *down;* the opening is too wide for the cat to straddle, otherwise.)

6. Finally, take away the diminished litter tin. Now it's between your cat and the toilet—and so many cats seem to love the arrangement!

If for some reason it's too much of a hassle for the cat, if it's causing great grief, abandon the plan. Some cats just never learn it—but *most* do wonderfully. No cat *ever* falls in—they just don't. I don't know why.

Other Solutions for the Uncouth Cat

Okay, you've tried everything. Your cat is still fouling the house occasionally, and breaking things and acting generally clumsy. You will have to assume that he is really trying to get your attention because he's in need of love, privacy or consolation. Perhaps there's a new cat in the neighborhood and he's aware of its presence as it patrols by the window every day—it's maddening for him. Close the blinds. If you can determine that there's a building site or any other unusual noise-making situation that could throw him, try to remove the cat from the source of the disturbance or buy one of those noise muters available in fancy gadget stores—you know, the kind that makes a low, soothing sea noise that masks sharper, more disconcerting sounds, or turn on the radio. How about some new toys to redirect his interest?

Finally, make sure you provide your cat with all the physical assurances, the love talk, the nuzzling, the stroking that you have time for. So many cats are desperate for attention, and this, plus emotional stresses you're not even aware of, can cause it to be *unwillingly* uncouth.

Problem Personality Three: The Crybaby

First, you want to hold it—it's so sad. That sweet cat is crying its poor little heart out. Then, you want to kill it! It's so maddening. That cat won't stop crying. You've eliminated the physical causes, your vet's checked it over thoroughly, it's not blocked or hurt in any way, as far as you can see. And you haven't had a decent night's sleep in a week. In *three* weeks. What in the world can you do?

The Crybaby Cat—Is It Lonely and Bored?

Perhaps you've moved to a new state. You know no one. So you get a cat. Or you've recently become divorced. You're lonely. So you get a cat. Or your children go to college, or your husband goes on a long business trip, or you've lost your job. You get a cat. Then you start to make friends. You test your own territorial waters. You begin to be secretly glad your kids are away—it means greater freedom for you. You get a fabulous job. Are you happy? Yes. Is the cat happy? No. The cat, accustomed to your companionship, loving and attention is suddenly bereft. You start spending whole days, whole *nights* away. The cat has separation-angst. It has been abandoned. It cries and cries and cries.

Remedy: Get Another Cat. A *perfect* solution! You'll be able to conduct your own life without feeling guilty for abandoning the cat if she has a playmate. Lonesomeness flies out the window. Naturally, introduce the new cat slowly and cautiously, giving the first cat the largest share of attention ... or else you will compound her troubles: not only are you leaving her, but you've brought in a competitor—God, that's *all* she needs, she'll figure. If you introduce her to a new pal, with discretion, though, *your* troubles may be over.

Remedy: Hernando's Hideaway. Cats have lovely private selves and they need private places in which to indulge them. There ought to be safe, secure hideaways (cleverly isolated from *your* bedroom) for your cat to dream in—a place where she can feel cozy, protected and alone. A place where she can think about her secrets. A box is an ideal spot to take the sadness out of a crybaby's life, and a box stocked with her favorite toy, blanket, maybe old shoe is even better. Some cats find their own spot under a couch or chair or in a closet that's always left open, but if your cat can't seem to discover her own best hiding place, provide one. I will often put a large paper bag *in* a box to provide an even more secret hideaway for cats who are experiencing stress. They seem to derive such strength from these womblike corners. Show the hideaway to your cat by bringing her over and depositing her in it. Give her a treat when you introduce her to her new, dark cave so she'll associate the spot with niceness. Sometimes, all a lonely cat needs for consolation is a darkish, warmish, secret place.

Remedy: Her Own Bed. What if your cat is only happy sleeping in your bed, *with* you, and you don't love the company? Set up a small chair right near your bed and make it invitingly cozy with a plush blanket. Show the cat her bed and whenever she jumps on *your* bed for the next few nights, gently but firmly deposit her back on the chair. She'll get the idea—and will still be close enough to you for security.

Remedy: Exercise. You may have an insomniac cat who cries all night because she's simply not tired enough to sleep through the night. I suggest evening play—keeping her busy for at least three hours before you retire by making sure you play with her or provide toys (which are changed fairly regularly) so that she can occupy herself. Ping-Pong balls are wonderful toys, by the way, and far better and more fun than any silly, commercially constructed toy. A Ping-Pong ball *in* a paper bag is the living end in cat fun! Try playing "throw-it-and-fetch" with a piece of crumpled silver foil (foil makes the *best* noises for her). If you have a backyard, but are afraid to let the cats loose because they've been declawed, try a tiny harness attached to a very long, lightweight rope. Always make sure it's attached securely so the cat won't eat it. Now, the cat can play while you read. Don't use a collar instead of a harness. Not only do cats seem to easily slip out of collars, they can easily strangle themselves if the top gets caught on a tree and they have to hang there from the neck. Once you establish a schedule, you're home free, and the cat will find ways to occupy herself until you and she are ready for bed. Exercise is the name of the game here, to work off her energy.

Remedy: Heat for the Natural Cuddler. Your cat has a natural affinity for heat—all those tabbies sunning themselves on windowsills are not doing it to get a tan. Why not use heat as a training device? Think of what it feels like for you to bundle up in bed with your electric blanket—delicious, nurturing, comforting. Now, electric blankets for cats (or heating pads) can be dangerous because if the cat's natural kneading motions rip through the insulation, exposed wires can present a threat. But a heat-moist pack, wrapped in a towel, lasts for hours— long enough to soothe your cat into complacency. Put it near his sleeping quarters *before* he starts to cry piteously: if you give him something nice in *response* to a cry, you're looking for trouble. So many of my clients train their cats to cry for attention, without meaning to. Gradually, as your cat is calmed through much of the night by the heat and gets into the habit of not crying, you can begin to remove the heat source so you're not married to it for all of the cat's life.

Warning: Whatever you do, don't get yourself involved in the Meow Means Gratification syndrome. Cats are nocturnal animals who naturally wander the house at night. If she's used to attention at 3:00 A.M., because she gets bored, figures she's up—everyone else might as well be, you're in trouble. Responding to whining cats can cause grief. Cut the habit short, immediately, with a strict "No" and a spritz from a water or air gun or gentle plant atomizer.

Remedy: A Doll's Bottle. The Crybaby is often a kneader and a sucker. Cats who are weaned too early or too aggressively often knead their owner's knees incessantly. This can really be a drag because not only is it annoying it can do great damage to clothes, stockings and bare legs. (Kneading stems from the suckling stage in a kitten's life and seems to encourage milk production in the mother.) For a week or so, place a piece of newspaper or aluminum foil on your lap every time you invite your cat to climb on to it. Kneading newspaper just doesn't provide that maternal satisfaction for most cats and often discourages the habit. You can also try pressing the soft pad under the paw every time your cat begins to knead.

Sucking wool may not be so annoying at first, but it can be a serious problem for cats who develop fur balls (or hair balls, as they're often called). Excessive grooming also leads to hair balls and, in turn, to vomiting. Often, a kitten or cat who has this problem feels lonely or attention-deprived, and warmth and cuddling go far—as does daily grooming to brush away excessive hair. A last resort: A cat that's sucking wool because it has been weaned too early may have to be encouraged to drink warm milk from a bottle for a while, causing it to revert to the sucking instinct. Then, after its fill of sucking, you can rewean the cat properly. (A doll's bottle with the opening in the nipple enlarged slightly is just the right size for this training procedure.)

Note: Wool sucking may also stem from nutritional deficiencies. Check your cat's diet with your doctor.

Remedy: People-watching. If your cat is lonely because you're away all day and it has no pals for companionship—a window to watch the passing world is always second best; also, there's nothing nicer than sunning for most kittens. If you have no windowsill on which your cat can perch—create one by putting a table or other furniture in the place a windowsill would be (keeping the windows closed, of course). And, although it seems superfluous to mention, keep the blinds *up* when you're out; so many people forget that and close off their pets from the satisfying pastime of people-watching. There are many reasons why a young cat cries, but the most common is loneliness and boredom. You can alleviate both by providing toys and a "movable feast" out the window while you're gone.

Problem Personality Four: The Destroyer

This is the guy who claws the velvet couch, scratches up the new, oriental Kirman rug, leaves claw marks on the dining room table and decapitates the geraniums.

Clawing

Here are some no-question-about-it facts:

• If you yell at your cat for scratching the couch, and don't train it to scratch on anything else, you will *never* get it to leave that couch alone. The best you can hope for is that the cat will cleverly stop scratching when she sees you coming. That's not too fabulous because she starts up again the moment you leave.

• If you have strong feelings about declawing cats, you'd better have equally strong feelings about not minding the destruction of your furniture or rugs.

A few words about declawing. I do discourage it, if possible, but if the choice is euthanasia and no cat at all—then, declawing the front paws is a more than reasonable alternative. Give your kitten or cat a chance to learn about scratching posts and other alternatives to furniture. Don't expect him to learn everything on the first day you've got him home. Before you hie him to the doctor, give him a reasonable shot at learning to live with his claws in your home.

But, if you cannot train him to use an alternative to your velvet sofa, my feeling is this: *No cat ever got a complex from having his front claws removed.* (It's rarely ever necessary to remove the back claws.) As a matter of fact, most cats will continue the clawing motion, if they're declawed in their kittenhood, which tells me they don't even know those claws are missing and, therefore, will hardly ever hold a grudge against you.

Clawing serves a purpose for the cat. If he's scratching up your home, he's not doing it to punish or spite you, but because some atavistic command, deep in his psyche, says—*Claw!* The cat, in the wild, uses his claws for defense or flight, and scratching them on woolly or rough surfaces has a distinct purpose: it pulls off the outer claws, which have frayed, to expose a new, sharper claw that's underneath. Clawing at rough surfaces also serves to keep the spring in cats' legs by giving its leg muscles a workout. All in all, it's a natural and important movement for cats.

But, if you cannot train your cats away from the furniture, and you care—declawing is not an anticat act, despite what some purists would have you believe. The cat might even, as I've said, continue the clawing motion, and if not, will hardly miss the claws if it's an indoor cat and doesn't need to fight or spring all that much.

Although many clients allow their declawed cats outside in prescribed areas, for me that's a no-no. Cats can escape from your safe haven, and although they can still climb without front claws, they can't climb so well or fast and might meet up with more than an adventure— indeed, distinctly hostile forces are in the outside world. Unarmed, they haven't a fair shot at survival.

Okay, you've decided to try to train the cat away from the furniture before you declaw. Here's how to go about it.

Remedy: The Scratching Post. Decide on the item that's to be used as a scratching post. Many people assign a particular piece of furniture the cat has already shown some preference for to be their cat's own property. If this is the route you go, determine in your own mind that you will not share his chair. It wouldn't be cricket to usurp the cat's scratching post even some of the time. Whatever you choose, make sure it's stable and big. Cats love to stretch out full length, and pull their claws along the item. If you're cheap and offer a niggling little post, or one that wobbles, don't be surprised if your cat goes back to the rug in disgust.

Some suggestions as to materials. I've never been one to believe that a cat will gravitate toward a nubby material faster than a smooth, silky one. What does seem to matter is that the threads of the material are vertical rather than horizontal; it seems to me that such material makes it easier to discard a worn claw sheath. Sisal is seductive to many cats. *Watch* the cat when you first bring it home to see what surfaces *it* seems to prefer. Then, take a similar fabric, tape it on to a wall or solid box (which you secure to the floor . . . if it moves, forget it—your cat will never approach it again). You might even offer a sturdy log with its bark still intact as a post for the cat. If you do choose a bark log, you'll have to replace it when the bark has worn off or else cover it with a material your cat will think is divine. The commercial scratching posts are almost invariably the wrong size, material and bulk—and they tip. Bad news. (An exception is the "Felix" brand.) The cat has to be able to sit on top of the post and scratch it from above—or lie out full length, and scratch from the floor. Some people try to make their posts beautiful—to match their home decor. I've even read a book that told cat owners that scratching posts can be "attractive additions" to their homes. Lots of luck. Of all the nonsense I've ever heard, that takes the cake. Don't, please don't try to make your cat's scratching post a unique and elegant addition to your home. It is what it is, and that *isn't* decor.

How do you get the cat *to* the post? Simple. Twiddle a little catnip mouse, on a thick string that the cat can't eat, along the floor, and let it "crawl" up the post. If you simply place the cat in front of the post, he and it will stare at each other with noncomprehension. Take the cat's front paws and rub them along the material to give him the idea. Try rubbing some catnip right onto the post. Try affixing the catnip mouse on top of the post— as kind of an ever-ready victim. If your scratching post is appealingly seductive, your cat won't ever need to destroy your furniture. *If he dares try . . .* get the atomizer ready. One spritz every time he's caught in the act of sharpening claws anywhere but on his post is a fine discouragement.

Make sure you place the post near where your cat generally sleeps because many cats tend to claw when they wake up—kind of a stretching reflex. Finally, *effusively* praise a cat who is scratching in the right place: "You *BRILLIANT ANIMAL!*" It helps enormously.

What about other types of nasty behavior problems to which the Destroyer Cat seems partial?

Stay Away from the Plants, Garfield!

There are a number of ways to discourage a cat from invading your geraniums.

Remedy: His Own Garden. The first thing you might try is letting him grow his own. That means growing *anything* in a pile of dirt and letting the cat go to town therein. Catnip is a superb herb to grow for your cat and will act as an aphrodisiac, making your cat a very euphoric creature—imagine having your own stash (legal, too) available.

Or, you can grow a variety of plants, among which wheat and parsley are popular. An easy cat garden consists of oats: sprinkle them onto a layer of good soil, cover the soil, give it water and sun, and in a week you should have an oat sprout salad for Garfield. Certain plants and flowers are dangerous for cats. Here are some of them:

Azaleas
Bulbs (most kinds)
Buttercups
Chrysanthemums
Crocus
Holly
Ivy
Lily of the valley
Marigold
Mistletoe

Philodendron
Potato
Poinsettias
Rhubarb
Seed pits (almond, apple, apricot, cherry, peach, pear, plum)
Sweet pea
Wisteria

As a hint to the cat that he'd be better off on his own greenery, try the following substances on the earth surface of your own plants:

• Aluminum foil (it's no *fun* to claw around on foil).

• Screens (the cat's claws, back or front, get stuck in the screens and cats hate that imprisoning sensation). Make sure you're around for this training—you want to watch that the cat doesn't get stuck for very long because he might injure his claws while attempting to pull them out of the tiny holes.

• Double-faced Scotch tape. An *enormously* effective deterrent.

• Balloons: One pop does the trick and even the motion of many balloons taped to the plant is discouraging to Tabitha.

• Salt or cayenne pepper sprinkled in the dirt.

• Pepper-loaded water: spray it on the plant.

• Perfume, Tabasco sauce or vinegar on the rims of the flower pots (and even on the soil).

• Decorative pebbles on the soil.

Jumping on the Furniture

Booby-trapping chairs and tables by propping a pile of books, paper cartons or cans so that they cascade down on the cat at the slightest touch is very effective.

One client of mine filled six low cookie tins with water and placed them all over the surface of the dining room table. Her cat soon got the message that leaping into a bath was not his idea of swell.

Cats, I've discovered, hate the smell of Jean Naté After Bath Lotion. Spray it where you're having cat trespassing problems.

Another client taped orange peels (the smell of orange is repelling to most cats) to a table top and that worked for her.

Still another client taped many balloons (blown up to the hilt) to couches and chairs for a week or so. Her cat hated the swaying movement of the balloons and their unstable nature when they were landed upon. The cat also did not appreciate the balloons' propensity to pop.

Chewing on the Lamp Cords

Still other cats think it's swell to chew up lamp cords— a potentially lethal habit. A bit of white vinegar or Tabasco sauce spread along a chewable attraction does wonders to change a cat's mind. It's a good idea to let him taste or smell the vinegar or Tabasco first—make sure it's as hateful to him as it should be—and then smear it on the electric cords or whatever you wish him to stay away from. If nothing helps, when you're away from home—disconnect! (or get track lighting).

Climbing onto Forbidden Shelves— When You're Not Around

Say you don't want your cat to climb to the top of the bookcase where you keep your precious antique books or bibelots. The cat is perfectly aware it shouldn't—and indeed it doesn't—invade this territory when you're around. The minute you step out of the door—she's on the shelf. *Whomp!* I have yet to see a cat not cured by an upside-down mousetrap. Any activity will trigger it off, not trapping the cat but scaring it totally. A temporary cure, that is. After about a week or so, the cat might venture back. Periodically set upside-down traps in areas you wish cats to avoid—just as a memory enforcer.

Don't hesitate to make your bad cat into a good cat. It is not mean or antifeline to be assertive in changing an animal's bad habits. I am certain that your relationship with that animal is cemented even more solidly when the cat respects your needs, as you respect its needs.

Problem Personality Five: The Prima Donna

Finicky, Morris-the-Cat types are not born, they're made. The Prima Donna whines when you're eating in a ploy to get some people-food, but refuses her own food. The Prima Donna has another tactic to get people to fall all over her with bribes—she refuses attention as well as food. The Prima Donna is a snob and totally spoiled. Here are some interesting points to give you ammunition in dealing with the Prima Donna. Studies have shown:

• Most cats (95 percent of them) will eat a new diet within three days or less if you make sure that absolutely nothing else is available.

• Fasting for several days is not at all detrimental and, in fact, can be a healthy boon as long as water is always available.

• You can tell that Morris is being finicky and not exhibiting illness from his refusal to eat if he scratches all around the dish: that indicates Morris is trying to "bury" his food because he thinks he deserves better. Sometimes a cat will drop a piece of food in his water dish—that also means he disapproves of your offering.

Training Remedies for the Prima Donna

Remedy: Don't Feed Her from the Table. Probably the worst thing you can do, the act that you'll regret for the rest of your life, is to feed the cat from the table. If Rusty has a taste for lobster salad, and you indulge it, you're sunk. First of all, you should be giving your cat a nutritionally sound diet, either from commercially pre-pared food or food you prepare yourself. (More on this in chapter 2.) Extra food from your leftovers creates imbalances of diet (not to mention pain-in-the-neck cats) because they never again want their animal food—only your prime ribs. It also makes mealtime (your mealtime) a drag with the cat either meowing and waiting to pounce, or giving you what behaviorist Paul Gallico calls the "silent meow" treatment. Patricia Moyes, in her book *How to Talk to Your Cat* says, "The silent miaow is exactly what it sounds like: the mouth opens, the head goes all the way back, all the gestures of mewing are there, but no sound emerges." The purpose of the silent meow, from the cat's point of view, is to make the human feel guilty. "See," it seems to say, "aren't I terrific for not making noise? Surely, you'll slip me that nice turkey slice, just this one time, for being so thoughtful." The silent meow is an insidious tactic to make one feel sorry for a starving feline, and, I'm telling you, if you fall for it, you deserve what you get—a Prima Donna.

Because food is your primary training tool (a treat for every success), you won't want to waste the impact by feeding treats indiscriminately from the table. Catering to a vocal or silent meow cuts the pins from under your training efforts by taking away the extraordinary value of a good treat—apart from regular meals. Cats learn lessons best when they're hungry. How do you expect your Prima Donna to pay attention to toilet training when she's filled up to her whiskers from your dinner leftovers? The *only* time your cat should be fed from your fingers (all other times from her bowl) is when you are pleased with her response—and want her to know it. Try feeding your cat her meals *before* you eat yours; that also cuts down on the nag, nag, nag at the table.

Remedy: Help Her Blow Her Nose. Cats appreciate their food through their sense of smell, like all animals—including humans. (Salt and sweet, bitter and sour are the range of our actual *tasting* powers.) If your cat has an invisible cold and a stuffy nose, unbeknownst to you, he just may lose his appetite because he can't smell the food. That isn't a true Prima Donna food refusal, of course, and your veterinarian ought to check out a recalcitrant eater's nasal passages.

Remedy: Vary Her Diet. Some cats have their owners running through hoops in fear that their cat will starve. Sample situation: Your cat has always eaten a particular brand. But you're away in the country, and the supermarkets there don't carry that brand. The cat's being maddeningly stubborn and hasn't eaten in three days. What to do? Well, first of all, when you get a new kitten, make sure you vary its brand of diet. If you always want to feed commercially prepared food for convenience' sake, alternate between several brands, all of which are certified on the can label by independent government agencies (the label should say NRC—National Research Council) to be nutritionally complete and balanced. Your cat will reject new foods, *of course,* if he's never been offered variety. Every now and then, you might cook a meal for your cat (see recipes on pages 27, 28, and 31–32). While we're speaking of food, incidentally, I quite firmly believe that moist, commercially prepared foods are quite superior to the dry. Dry foods are high in ash (they cause cystitis and bladder stones) and too low in protein. Cats need a great deal of moisture in their diets to facilitate urination and dry food simply doesn't have any. Sure, the dry food may stimulate teeth and gums, but this can be accomplished through treats.

There is one exception to my dry food rule: On occasion, if a cat is prone to having chronic constipation, or soft stools or diarrhea, all of which should never be ignored, I recommend a certain dry food called IAMS cat food, which, in my opinion, is the best of the dry foods because it is low in magnesium, a mineral that predisposes cats to calculi formation. Calculi are urinary crystals that can cause cystitis in male and female cats and cause urinary blockages in male cats. If a food is high in magnesium (over 1 percent) many cats will develop these potentially fatal problems. IAMS is nutritionally balanced and contains less than .1 percent magnesium and a calcium/phosphorus ratio of 1.2 to 1, safely within bounds. Remember that a cat food may advertise that it's low in total ash but *still* it may be high in magnesium. It could be low in total ash because it has insufficient levels of calcium, phosphorus and trace minerals—all of which your cat needs for total health. So, if your cat seems to be having soft stools or constipation, do give IAMS a try, being particularly careful to provide clean, fresh water at all times.

If you have no such problems with your cat, your best bet is definitely a commercial, moist, canned food that is low in magnesium and ash and which has an adequate calcium/phosphorus ratio.

You needn't worry that the moist, canned food will spoil too easily, because fat stabilizers keep it preserved for a long while. If you are a believer in natural foods, and don't wish to cook for your cat, there are many brands that are holistically sound and also moist— Cornucopia is a good choice, for example. With these foods that are all natural, there are drawbacks: they must be carefully refrigerated because there are no preserving additives; and they are quite expensive. Cats, incidentally, far prefer freshly opened cans of food to those that have been stashed in the refrigerator for days.

Remedy: Camouflage Her New Food. If your cat refuses to eat new foods, mix a little of the food she best loves into her new offering—even if it is a treatlike food; because the smell of food is the big attraction with felines, the older, more familiar food might induce her to be more experimental. Feed the cat the mixture, gradually increasing the new food and decreasing the old for about four to six days until only the new food is being offered—and eaten. Warming the new food often does the trick.

Caution: Do not offer anything other than the old-new food mixture during the four to six days of training.

Remedy: Harden Your Heart. The hardest thing about a Morris-the-Cat type is the way he looks at you when he refuses his food ... "Hey, gimme a break, will you? ... This food is not fit for *mice!*" If you want your cat to start eating a more varied diet, and you know, in your heart of hearts, it's you that is sabotaging it by offering a favorite food as soon as your cat rejects a new food—feed the cat, or change his food, when you're going to be *away* from home. Don't torture yourself by wondering if you're doing the right thing. You are. No healthy cat ever starved when food was available. Let the cat be a Prima Donna—*away* from your vision. It'll do wonders to convince him that eating is nice.

Remedy: Clear That Food Away! One of the primary causes of finicky cat behavior is leaving the cat's food available and out all day long. First of all, the smell of the food, rather than the taste, is what turns cats on. When food is smelled, the cat's body prepares itself for digestion, and body metabolism is considerably slowed. Blood supply to all the organs except the stomach is undersupplied during this preparation. When food is constantly available, the body doesn't have a chance to metabolize normally and the cat's organs are undersupplied with blood too much of the day. Result? Impaired circulation. Poor circulation causes innumerable physiological problems which in turn makes a cat lose his appetite.

Leaving food perpetually available *creates* finicky eaters. It's a good idea to leave food out for a half hour, *at most,* morning and evening feedings. (Water is, of course, always available.) Wash your cat's dish thoroughly so no smell remains to trigger off the digestive process. Don't worry if the cat doesn't eat for a day or two—in fact, cheer if that happens every now and then. A periodic fast can never hurt; it *helps* the body to purify itself. I once treated a cat locked in an airline storage locker by accident, for a month, while in transit to another state (the airline carrier had "misplaced" her much to her owner's frantic dismay). The cat was finally found, extremely dehydrated, starving, but quite functional.

Remedy: Skip Lunch. Grown cats eat *no more* than twice a day. If you supply lunch, don't expect a hearty dinner appetite.

Remedy: Smell Swell. A powder called Kyolic (aged garlic), which can be bought in health food stores, is usually irresistible to cats when sprinkled on food.

You might also try brewer's yeast, onion powder or cod-liver oil on food to add zip to its appeal.

Remedy: For Whom the Dinner Bell Tolls. Feed your cat on time. The cat is a creature of habit, perhaps more than any other animal, and she thrives and dotes on routine. Try feeding her at the same time, every day, to *set the memory* that *now* is the time to eat. Charles Henry Ross, in his *Book of Cats,* written in 1868, tells a story that punctuates this point. There was a cat who lived in a monastery, he recounts, and he was fed every day just after the bell tolled to call the monks to dinner. Shut away in a room accidentally, one day, he missed his dinner when the bell rang. When the monks had dined, they searched for the missing cat and let him out of the room. A minute later, they unaccountably heard the dinner bell ringing for the second time that evening and when they went to see why—there was that aggrieved and hungry cat, hanging on the rope and tolling the bell for *his* dinner. (Thanks to Leonore Fleischer in *The Cat's Pajamas* for this tale.)

Remedy: Hug and Kiss! What about the cat who refuses attention, as well as food—the finicky-about-people cat. You may *never* make him the life of the party if his kittenhood wasn't filled with toys and games and much loving. With patience, though, and with a great deal of game-playing and stroking and conversation, you can take a cold, withdrawn cat and make him quite a bit more affectionate. Choose the mellow hours to touch and hold him—after you've both had your lunches, when he's soothed by a stray beam of sunshine floating through the window. *Insist* upon contact if your cat looks uninterested. Studies have shown that severely disturbed mental patients eventually do respond to loving attention. Other studies have shown that the same mental patients do wonderfully when someone gets them a pet—that miraculous ability of touch to heal psyches. Perhaps you might consider getting your cat a pet for the same reason. Sullen, finicky cats with puppies or kittens to play with are often found to melt with love.

And that's the way to deal with the Big Five Problem Personalities.

Let's assume, now, that you have a reasonably compliant cat—one that's not a brat to have around. Chapter 4 will give you the training techniques—those methods to use to best train a cat *to* do something—not to stop doing it. These are the most gratifying and fun-filled types of training. Read on!

4

Training Your Cat to Start Doing Something

C'mon Clarence—You Can Do It!

Live with a cat on *his* terms? No way!

It's funny—the people who are absolutely dogmatic in their views, who bend everyone else's will to their own, who call the shots in each of their personal relationships—are most likely to let their cats walk all over them. Despite what you might think, a cat's tread is *solid*. Carl Sandburg wrote about the fog coming on "little cat feet" to imply soft, hazy, ephemeral substances, but a cat that's been crowned Queen of the House because her owner hesitates to impinge on her sovereignty is a cat with a heavy tread—almost intimidating. It's very difficult to live sweetly with an untrained cat. They begin to act like the wild creatures they were eons ago, instead of the comfortable, loving, independent-but-reasonable creatures into which they've evolved. Look: you'd find it difficult living with the prototype caveman or cavewoman; just as people were civilized, so cats must be with the use of love, reason and wit as training tools—not to mention the ubiquitous snack, without which no training can occur.

Those who would train their cats to do parlor tricks are of a certain breed of people to whom I lay no claim. It *is* possible to teach certain amusing tricks to make your cat a whole media event, and I'll talk a bit more of that briefly. Generally speaking, though, I belong to that club of humans who feel a cat's training should be limited to that which makes her a doll to have around.

Training your cat won't make her a robot, but will channel her behavior into patterns that make life more pleasant for you and for her. But behavior training relies on *positive* reinforcement. Negative reinforcement or punishment invariably fails.

To recap: we've spoken of three techniques to use in training cats to *stop* doing something:

1. *Manipulating the environment.* I've spoken of the Barnyard Effect: simply by cutting down on the number of cats in a household, one can often eliminate spraying or nasty aggressiveness. The presence of another cat in heat can drive your cat to drink, even if the aggravating cat is outside, and simply by removing the sight of this parading cat by drawing the blinds over your own windows can solve the problem. By changing the material in the litter box to one more appealing to your cat, you can "train" him to use it rather than the bathroom floor. In these ways, you manipulate the environment.

2. *Changing the physical state of the cat* to produce "training" effects. Declawing, spaying, even using tranquilizers can change behavior problems. (Regarding declawing—it's always a *last resort* with me. I try to discourage it on principle, but before I'll see a cat destroyed or abandoned, I'll always recommend it.)

3. *Making sure your cat is medically healthy* also

clears up behavior problems. Cystitis, blockages and diabetes, as I've said, are just a few of the diseases that can alter urination patterns. Metabolic, neurological and structural problems (like askew intervertebral discs) can cause obnoxious behavioral patterns. When a cat hurts, it often responds by crankiness, as many of us do.

And last, there are the *behavior modification techniques*. These are the ways to encourage a cat *to do* something. Once you have a healthy cat, orchestrating new behavior by using these training techniques makes it a wonderfully desirable cat.

Par for the Training Course

Before we get into the actual training techniques, let me tell you about my *PAR-for-the-course theory*. There are three tricks to training, three words that make the difference between success and failure. They are

1. *P*ersistence
2. *A*ssertiveness   —In a word: *PAR*
3. *R*epetition

1. *Persistence.* When training cats, it's important to know that you will succeed after a million tries (or what feels like a million tries, anyway). Your cat has to know that you won't forget about the thing you're asking him to do even if he pretends that he hasn't the foggiest idea of what you're driving at. *Persistence* in training is the key to getting the cat trained. For that matter, cats don't understand it when you're not consistent *as well as* persistent, either. They don't see the fine points of *sometimes, yes, sometimes, no* and *maybe* or *hardly ever*. They know "good cat." They know "no." They know

black and white. No gray. No maybes. You have to be patient and you have to give time to training. How much time? Depends on you. If you're home and can work with the cat off and on all day for a couple of weeks—it should take a couple of weeks. If you come home from the office and work with the cat for a half hour a day, it should take at least a month or two for him to learn a trick well.

2. *Assertiveness* is the second trait you must display. You can't ask or beg your cat to sit, or stay, or come. You must *tell* it firmly, in no uncertain terms, that you're the boss here, see, and what you say goes. All this can be done in quiet, firm tones—no yelling or harshness is called for in assertiveness. Your attitude is by far the most important key in training. Positive and assertive demands come across with power. Cats are strongly keyed into your moods and doubts and if you think you will never teach the cat to do something, she will catch your indecisiveness immediately. Far better to impress upon the cat, by your firmly assertive nature, that it is to her infinite advantage to abandon the idea of lolling around on your bed.

3. *Repetition* is the last turn of the training key. Do it over and over and over. When your cat has learned a method, never take it for granted (as you might with a dog) that she'll know it forever. After a few weeks of not being pressed to perform, cats tend to pretend they don't know anything about it at all when finally asked again to show off. To keep their performance reliable, cats need periodic reinforcement of learning. That means that even if you have no need to call your cat to come for several weeks, do it anyway, just to keep her on her toes. In the actual training, repetition counts greatly; the repetition of the learned act all through the cat's life also counts.

Remember ... *P-A-R for the course*—Persistence, Assertiveness and Repetition: the key to training your cat to do something he wouldn't do on his own.

No One's a Saint

There's one more thing you must know about training and that is it's unfair to expect absolute virtue from a person—let alone a cat. It you train a cat to stay off the dining room table, and then leave the house with the remains of a tuna fish sandwich on that table, don't be shocked off your feet if the cat jumps on the table to finish off the snack. There are reasonable limitations to what you can expect from a cat. You can train her to go into a litter box—but if you then give her an abundance of heavy cream or something else that may cause diarrhea, don't be surprised if she has an accident outside of the box. You can train her to come when called, and she can do it *most* of the time, but if she's outside and there's a handsome Tom prowling around, spayed or not spayed she just may not heed your call. No one's a saint.

Last Wise Words

Remember—the "no" word is only used in cease-and-desist maneuvers when you're telling your cat to cut something out of his behavior that you hate. No! to biting the baby, No! to jumping on the table, and No! to scratching the furniture. In training to do something, *the positive approach* is far more effective *and* relaxing to you and your cat. If the cat misses the jump, you don't say No! but you *do* reward her when she makes the jump, and you do say, "Good cat, smart cat," in your nicest tone of voice. You can spend a month trying to train Melinda by yelling, "Wrong—Oh, dumb-o—God, you are a *stupid cat* . . . No! *Don't* move when you're supposed to *stay!*" And, at the end of the month, you will be no further along than you were when you started. Or, you could pepper your training with *positive* and encouraging words, "Oh, yes . . . Oh, good kitty, you did it—here's a nice piece of liver!" and have the most well-trained cat in town.

The Fabulous Four-Letter Word

It's got to be in every cat trainer's vocabulary. The most potent training tool without which I believe training cannot take place is *food.*

As much as you'd like to believe your cat responds because she loves and admires you, it's not true. Cats think about, dream about, are just crazy about food. It's their modus operandi, their Kama Sutra, their Pledge of Allegiance. A pocketful of dried cheese or liver bits or some tiny, succulent yeast tablets are to a cat what two hundred mackerel are to a trained seal. We use food to spur interest, and without an implied treat it's doubtful that you could get a cat to do anything he didn't think of by himself.

The Training Techniques

How to Teach Your Cat to Come When Called

Apart from being helpful (you shouldn't have to track him down at feeding time), having a cat come when called can be a lifesaver. Cats that have been declawed, and then escape into the real world of other cats, dogs, cars and danger traps have little hope of surviving for long. It's wonderfully comforting when you're searching for a lost cat to know that he'll respond to his name if you can get within distance of his hearing. If you prefer, you don't even have to use his name if you and he have another *sound* you share. The point is to design a special call, one that's unique to you and him, that means "come." I call it . . .

The Designer Summons

Decide what you want your cat to respond to. Will it be her name or a sound like a party clicker—or a combination of both? (It should never be an unpleasant sound like a whistle: cats have acute hearing and can catch sounds that are inaudible to us, so a whistle can be nothing short of torture.) If you choose a name to summon your cat, make it a short and sweet one so the cat can learn it easily. A client I had named her cat Mr. Short and Sassy, and then complained that the cat would not come when she called it. The cat was *embarrassed* by its name, I just know it. Remember that a cat will only answer to her name if a warm and loving relationship exists between her and her owner. Repeat her name constantly during playtime, feeding, petting and conversation. Be consistent. If you use just the name as a designer summons, never use the name *and* a sound. If you use both—a name and a clicker, always use both when you call. The point of a designer summons is that the call is designed especially and personally for the cat. If your name is Mo and they start calling you Buddy, it will confuse you.

Remember that cats also have an ability to funnel a sound directly into their range. Their ears are shaped in such a way as to zero in on specific sounds—the way you can make your hearing more acute by cupping your hands around your ears to pick up desired sounds. A cat that is trained to respond to a clicker will come running for miles toward the sound of that clicker.

As soon as your cat *is* your cat, begin the imprinting process; that is, teach him that you're his special person, his own. Since he does depend on you for his food and comfort, let him know it by giving a healthy dose of love and affection along with the essentials. Every time you put that kitten's food down, even if he's right in the same room with you, call him by using his name or the special designer summons you've decided.

"Come, Spot. Come, Spot." (I know a woman who named her cat Spot so she could say, "Out, out, damn Spot . . .") Never alter the call. Never alter the tone of voice, if possible.

"Come, Spot. Come, Spot." If you use a clicker or a clap or a melodic yodel as your designer summons, use it *every* time you feed the kitten. The key is consistency. Food is not to go down unless you've called the cat to come to it. When he responds by coming, praise him effusively. Knock yourself out by telling him (and stroking him) how clever a cat he is. Make sure you train your cat to a sound that is always available. One client thought she was brilliant because her cat always came to the sound of the electric can opener opening the can of cat food. She'd turn that opener on and the cat would come bounding. When one day the cat became lost on the beach during vacation, the electric can opener was of little use.

Gradually, begin to call your cat to come for dinner when he is in another room. Always reward him with dinner, when he comes. If he *doesn't* come, retrieve him, put him down near the food source, say "Come, Spot"— then put the food down.

Caution: Never call your cat for something unpleasant that's about to happen to him. You may unintentionally train him to run away instead of come when you call. A client of mine was in the habit of calling her cat to come to her to receive a verbal chastising every time the cat scratched her dining room chair. Sampson, the cat, soon learned that "come" meant bad news for him, so "come" was a command to Sampson to get under the bed, fast. If you want your cat for something he's going to find irritating, like having his nails clipped or being put in his carrier, go get him—never call him to you. Call him only for a food treat or a love session!

After a few weeks of this training, and success is making you both happy, start calling Spot (always with the same designer summons) when it's not mealtime. Call him at three in the afternoon, say, but when he comes always reward him with a small snack-treat. It's important to keep up the life of the snack-treat concept. Sometimes, you will be calling the cat when a snack's not available, and, of course, you can do this sporadically—*but not too often.* Your cat should always associate "come" with a bonanza for him. Keep this in mind: cats adore ritual. They love familiarity with the same procedures, the same routines, the pomp and ceremony of being appreciated. When you establish the ritual of calling your cat, rewarding it with a snack, praising it effusively for its response, Spot (or Tinker or Whoever) is in Cat's Heaven. If you can work it so that the ritual is always pleasant, always delicious, your cat feels safe, appreciated and eager to comply. If you call your cat (even just once)—and when it gets there, you yell at it or make the act of coming an unpleasant experience, forget convincing it to come when called. You can undo months of positive reinforcement by one negative act. On the other hand, if coming when called is rewarded lavishly—the cat will surely come.

Always feed your cat his treat with your hand so he gets the idea that *you* and *your* call are the source of goodness. Putting a snack on the floor is meaningless; the cat will think it's serendipitous that the snack got there, but will not associate *you* and *your* call and *your* voice with the treat.

Training Tips

• Work with the designer summons in a quiet place so that the distractions of a busy household don't interfere with a "come" command, at first, anyway. Never start your training outside—there are far too many distractions. *If* your cat will be allowed outside, eventually reinforce the designer summons outside. You will be amazed to find your cat, who has been trained to come inside first, leaping and bounding to get at its snack when you click your clicker or say your special sound outside.

• Don't ask your cat to repeat success again and again. She will become infinitely bored, as you might also. Once she's responded *well* and gotten her snack, don't call her again for at least an hour. Calling and giving a treat, and then repeating the process two or three times takes away from the joy of the treat.

• It's a good idea to hold your training sessions when your cat is hungry. Circus cats are always worked with or asked to perform *before* they're fed; it is then that they respond best to the "food" word. Incidentally, never give your cat the treat reward before or during the training exercise—only *after* he's correctly accomplished the chore at hand.

• Make sure you reinforce your designer summons with affection. Stroking is like an opiate to a cat: once he gets to love it, he can't do without it. When he hears his summons, he knows he's in for a treat and some touching—and both put him in ecstasy!

• Don't nag by repeating commands. If you say, "Come, come, come on, baby, come—," the cat will come only when you say, "Comecomecomeonbabycome." *One* "come" should do it.

Getting the Cat to Come,
Even When He's Been Scared

After you teach your cat to come under familiar circumstances and surroundings, also teach him to come when there's trauma or strange territory involved. This is particularly useful when the cat's outside by design or escape. Here's the technique:

In your controlled home environment, ask someone to make a loud clatter or other startling noise. When the cat has been alarmed (not scared out of his wits, please), call him with your designer summons. Keep doing it until the cat responds to the call. (Only call *after* the startling noise: if you call before, the cat will surely decide that a call from you means a big, unpleasant bang shortly thereafter, and will run away, not come, when you call.) If you have nobody to help you with this training, you can throw a folded newspaper or a tin cup to land somewhere near him (not to hit him) and then call; the cat should not associate you with the thrown item. Coming to you should, in other words, represent safety, food, cuddling, praise. Repeat, every now and then, without *severely* shocking the cat. To teach a cat to come when it's been startled or frightened may, one day, be of inestimable value.

Some Wise Words

When you call your outside cat in, check him carefully. Confusion has a way of coming in the door on little cat feet. A client of mine had her cat castrated by another veterinarian before she moved to the city where I practice.

She called her cat in one day, and after a while noticed that Sidney, previously the soul of affability, was getting kind of snappish. Snarling and biting, Sidney was acting like a strange cat when he was touched. She took him to the veterinarian to see if any illness or injury could be causing Sidney pain, and thus make him hostile. Sure enough, the veterinarian did find an abscess he wanted to treat. In making his examination, though, besides finding the abscess, the puzzled and excited doctor had discovered a first in medical history—something incredible. Sidney, quite unbelievably, had grown back his testicles.

"I'll treat the abscess," the doctor told my client, "and while Sidney's under anesthesia, I'll neuter him again. I want to study this case more thoroughly anyway." He'd never heard of the reappearing-testicles phenomenon and had visions of reporting the case in a prestigious medical journal. My client went home, after leaving Sidney for a few days so that the doctor might study him and write a report to the American Veterinary Medical Association.

Two days passed. During the night of the third day, my client heard a persistent mewling outside. She opened the door and—in walked Sidney. The *real* Sidney . . . who had disappeared when a look-alike cat walked into his home.

My client felt terrible about informing the wildly elated doctor that medical history had not been made, and she felt even more odd knowing that someone else's cat had been castrated in Sidney's name.

Make sure your cat is *your* cat.

Finally, if you've called and called all day with no success—try using the designer summons *at night* when there are fewer street noises and the cat might be more likely to hear your voice. A client of mine was once about to hire a private detective at an astronomical fee to find Raisin, her missing Burmese. She heeded my advice about night searching and saved herself the embarrassment of telling the detective his mission was to find a missing cat or self-destruct.

How to Teach Your Cat to Stay and Stop

Once your cat has learned to come, he ought to learn to stay (or stop). This is an important command because it may save his life.

In this command, the use of a hand signal will be important. As your cat advances, thrust out your hand with the palm held up at right angles to the arm. This is a message to the cat that you are making a "wall" against his advancing any farther. Naturally, your cat will have to have you in his vision first, to see the signal as you say firmly "Stay!" or "Stop!" (It doesn't matter what you say, as long as you say the same thing every time.)

At first, thrust your "wall" quite close to the cat, so he will have to stop. When he does, praise him happily and give him a treat. Eventually, drop the hand signal so the cat is responding to the sound and not the sight. This is quite important because you cannot always be directly in front of the cat when you wish him to stop advancing on the waxed floor, the route to the open door, and so on. I have seen cats stop on a dime when their owners say, from quite far away, "Stop!" The training takes repetition (a few minutes every day won't make the training sessions boring or tedious) and patience, but it works eventually. I promise that. Once you have taught your cat to stay and stop, you must teach it that it's okay to go again. This is called a *release*. Choose a word that will release the cat from the stay position—a word like "okay" or "free." Accompanying the release word with a sharp clap of the hands acts as a punctuation mark. It works like this:

"Stay," you say to the cat, and he stops dead. "Stay," you say again, as you back off, so you're a good few feet away. Then, "Okay," you say with a sharp clap. The clap impels the cat forward and when he moves from the stay position, again you praise him and treat him. Don't call his name or use other words because that will only serve to confuse. Just, "Stay" and then, as a release, "Okay." It is never really necessary to use the cat's name in this training. The training words alone will suffice and grab his attention, once they become familiar in meaning. The simpler you keep your commands, the easier it is for the cat to learn. If your cat breaks the sit or stay command, *before* you give the release word, just retrieve him and take him back to the spot. Go through the motions again. When you are certain (and this will take weeks, maybe months—patience is the key) that "Stay" and "Okay" are learned, try the cat out in a room where a distraction will occur. After all, you may well want him to sit just when another cat, or a dog or child enters the room. Practice, practice, practice. Gradually increase the time he has to stay. If he breaks the training, go through the routine again. Do not praise or give treats, *until* the cat has done what you wish. If you say, "Good cat," for simply what you feel is a good try, it will be quite confusing. Only real success should breed a treat and a hug—harsh as that sounds.

I might add that the commands are to be kept up. If you successfully teach your cat to stay—and then let five months go by without asking him to use his newfound knowledge, you'll have to start all over again. Cats' memories are not legendary; they need bolstering every once in a while.

How to Train an Outdoor Cat into Being an Indoor Cat

Here's a very common scenario: Toots was an outdoor cat. His master, a client of mine, put up with ear mites, worm infestations and fleas and ticks. He managed to survive BB gun shots, other cat and dog attacks, errant skunks and runaway cars, but the day a nasty urchin tied tin cans on Toots's back legs was the day my client decided that, henceforth, Toots would be an indoor cat. He'd heard (correctly) that being an indoor cat would add an average of five to seven years to cat life . . . and that was also an incentive.

Not so fast. Toots didn't buy the new plan. He cried at the door constantly because in the past he knew that crying impelled someone to open the door. When he wasn't crying at the door, he was just sitting there pitifully, breaking his owner's heart. Finally, in desperation, the two of them came to see me. I knew some ways that the outside cat could be converted into an inside cat, and offered a choice of alternatives:

• Every time Toots begins to walk toward the door, interrupt his walk by giving him a tiny treat or some milk in his plate—*away* from the door. This acts as a reward for *not* going to the door. Intercepting the cat's objective by offering treats and affection works very well after a while.

• Ignore all crying or mewing at the door—even if it's painful. Once Toots walks *away* from the door, he's to be praised and petted by anyone around.

• Give many new toys and create diversions during

the day to keep the cat occupied. Make sure that there's an indoor place for scratching and a place that climbing is permitted to replace the scratching and climbing outside. A commercial "cat tree" with many perches is a nice thing to try.

• Finally, if nothing works after a couple of weeks, use punishment. Keep a water or air sprayer near the door and every time the cat sits near the door or cries to be let out, spray him with some water or air. When he walks or runs away from the door, reward him.

Another effective punishment is Noise Therapy. Cats hate shrill noises—the worst of which is a whistle. One client blew her terrible whistle whenever the cat began nagging to go out. He studiously began to avoid the outside door. A footnote to this story is that the same client began getting a series of obscene phone calls and the police suggested she blow a whistle into the phone whenever the breather made "an appearance." It worked. Well, it worked on the breather, that is. The cat showed definite signs of hysteria. It was really confused. "Look, I'm nowhere *near* the door and still—that damnable sound." When my client realized what she was doing to her cat's nerves, she stopped the whistle, fortunately *after* it had discouraged the obscene calls.

There is no question that many studies have conclusively shown that indoor cats *can* evolve from outdoor cats when behavior modification techniques are used.

How to Train Your Cat to Sit Still for Medicine and Other Ministering

The word "train," in all honesty, may be inappropriate here. What we're really talking about are techniques and attitude. The attitude of the cat depends on your attitude:

if you're nervous and tense about clipping its nails, absolutely terrified you'll cut too much and invade the "quick," the cat will feel your insecurity and panic. On the other hand, if you hold her with a firm and practiced touch—a no-nonsense grasp that doesn't clutch, doesn't squeeze, just *holds*—your cat will respond in tolerant tractability. Let's first talk about medication.

Medicating Your Cat

If you have to administer pills, eye or nose drops, or have to force feed your cat, try this technique to protect both you and him:

1. Place a Turkish towel, small piece of rug or other plushy substance on a table facing you for the cat to dig his claws into. Place the cat on the rug-lined table.

2. Drape a largish towel over the cat's neck, making a "bib" around the cat to prevent him from scratching you.

3. Tilt your cat's head toward the ceiling by rotating it to the right until his nose is directed *to* the ceiling. This relaxes the chewing muscles and makes it easy to squeeze open his mouth with your other hand gently but firmly. Grease a pill or capsule with butter (whipped cream or cream cheese are other "slippery" things to try) and place it far back in this throat—over the little bump on the back of the tongue. Stroke his throat downward (a reflex action will cause swallowing) or blow on the cat's nose for the same effect. You know that he's swallowed the pill when he licks his nose after you've released him. For liquid medication, use a plastic (*never* a glass) medicine dropper and squeeze the liquid a bit at a time into the side pouch of the mouth between the gums and lips. Don't give too much at a time or the cat will choke.

Caution: Always hold your cat's head straight in front of you for liquid medication and never pointed up, because liquid can be inhaled into the trachea if the cat's head is facing the ceiling when he tries to swallow.

4. If the cat is *really* unruly, and at last resort, you can place him in a pillowcase or small laundry bag with only his head showing before administering the medicine. Cats do not appreciate this position, though. Would you?

Claw Clipping

A cat should be well manicured if it's not declawed, to protect both you and your furniture. Here's how to do it:

1. The best place for the job is on your lap. Arm that lap with a Turkish towel or rug in which the cat can dig the claws not being worked on. If it resists being held on its back, you can also try holding the cat under your armpit for warmth and security (the cat's, that is).

2. Place a not-too-tight rubber band around the cat's head and ears—yes, its *ears!* This works wonders to take his mind off his feet; or, have someone tap his nose or play in front of him for distraction.

3. Maneuver the cat on his back, if possible, so his feet are in the air. Press up gently on the pad under each nail to "let out" the nail from its sheath. Don't squeeze— just press *up* on the pad under the claw and the nail will come out.

4. Hold the paw firmly with the "let-out" nail extended. With a nail clipper designed especially for cats, clip off only the curved part of the claw. Work near a light so you can see the fragile vessels that run down the center of the nail *which you want to avoid.* If you accidentally cut into this pink part, the cat will yell and you'll see blood. Press a bit of cotton against this "quick" of the nail to stop the bleeding. My favorite blood-

stoppers on the claws are old-fashioned soil or a bit of sweet butter placed close to the bleeding part. If plant soil is used, be sure it's not adulterated by fertilizer. A styptic pencil can also be used, although that can sting a bit. Don't panic—paws are a long way from the heart.

5. *Hint:* Work slowly, reassuring the cat all the while that he's being extraordinarily brave. This should not be a traumatic experience for either of you. When you're finished, reward him with a treat so he comes to associate clipping with reward. Twice a month is plenty often for nail clipping.

6. *Second hint:* During your regular playtime, make sure you stroke and pet the cat's claws so he gets used to your handling them. Furthermore, this will give you a familiarity with the construction of the paw and how the nail slips in and out of its sheath.

How to Train Your Cat to Walk on a Leash

Most cats figure that leads and harnesses are for the dogs, but it's a good idea to train your cat to get used to a lead. When you're on a long trip, lead walking will be essential for breaks and exercise. If you have a big-city cat who spends time in the country, you'll need to give him a bit of air for his psyche. Lead walking does not have to be a matter of training *you* to walk at your cat's speed and whim. You can make it easier by doing it right.

If the cat's a skittish, nervous, shy type—lead walking simply won't work. He'll be too alarmed by every passing diversion, every falling leaf, every blaring car horn. Most well-adjusted cats, though, *can* be taught to enjoy a stroll outside and although they'll never stride along, heeling dutifully like Fido, they will learn to follow your sauntering pace.

First of all, use a figure-eight harness—not a collar. It's safer: the cat can't slip out of it or, if she escapes, get caught and hung by it. The leash should be of the lightest weight, cotton variety—definitely not a chain, which is far too heavy for a cat. Some trainers recommend fishing line for a lead—it looks as if the cat is not on a lead at all but trained to walk by your side—unfortunately, fishing line tends to coil and knot easily and is a pain in the neck, despite its show-off value. Make sure your lead is not too long—the idea is to train the cat to stay reasonably close by your side and a hundred feet of clothesline won't do it. A six-to-ten-foot standard lead is fine.

Get your cat used to the harness and the lead *before* you attempt to walk him on it. Let him sniff it, make friends with it, wear the harness around the house for a couple of weeks before you ever attach the lead to it. Soon, it becomes as familiar to the cat as his own fur; he's not even aware it's on.

Next, attach the lead to the harness but don't hold the other end in your hand. Just let the cat pull it along for a while. Compliment your feline profusely. Give him a treat. Do this intermittently for a day or so, till the cat gets used to the weight of the leash. While he's practicing inside, make sure it doesn't get tangled in anything, which will frighten him.

Next, pick up the other end of the leash. Go where he wants to go—still in the house. Don't attempt to lead him, at all. The idea is to get him used to *you* being on the other end of *him.* Make the walks short. Occasionally, give him a tiny tug and a sense of direction to the lead. If the cat resists, follow *him.* Tug gently when he lies down or if he simply refuses to move at all. Cajole him to move by holding out a treat a couple of feet from his nose (it helps to have a helper or twelve-foot arms). Do not give him a treat for lying down or being stubborn. Only give the message that *walking* on a lead is to be rewarded.

Gradually, take over as leader. Practice walking along, in the house, until the cat is following along. Always reward for a good performance. Don't make the training sessions too long but do try to incorporate two or three sessions during the day as memory reinforcements.

Never *drag* your cat along because he will hate the whole experience and write leash-walking off his list completely. Stick to a straight path and gently tug when the cat stops to sniff. Never follow him, now; you're the boss at this stage. If he refuses to move, wait him out, all the while offering a treat just ahead of his nose, talking softly and encouragingly.

The moment of truth comes when you and the cat must face the great outdoors. Understand that it can be a frightening, even terrifying experience for a cat who's never seen sky except from inside a carrier, peeking out. Be prepared to stroke and hold the cat a lot. Allow him to sit, if he wishes, or walk—with you following—in this initial getting-used-to-the-outside. His first walks should be in a relatively calm place, preferably very early in the morning when few people are about. Say your cat's name over and over, as in "Come along, Kipling. That's a good boy." Make the first walks short and repeat them for a few days.

After about a week, you're ready to encourage the cat to walk *along* with you. It takes time and patience. Remember, never *drag* a cat. He knows what you want— haven't you had those in-house sessions? He should be able to transfer the training to outdoors once the enormity of the place is no longer a threat. Patience and consistency are a must.

●Don't hesitate to pull on the cat's leash, if necessary—it does *not* hurt, despite his best efforts to make you think it does.

●Don't expect him to march along, heeling like a dog. Can't be done. Never was done.

●Be prepared to lift him up when an aggressive-looking dog or kid passes, but don't swoop down on him constantly, or you'll make him a nervous wreck.

●Don't walk your cat when you have to get somewhere in a hurry: it won't work.

●Your aim is to make your cat feel secure and comfortable on a lead and to give him a taste of the larger world as you become an outdoor team.

●External stresses of people, movement and noise can be decreased by exposing the cat to the outdoors slowly, for increasing time periods.

How to Train Your Cat to Use a Cat Flap

If your cat is both an indoor and outdoor cat, a cat flap is a wonderful idea that releases you from being a victim of the cat's cries to be let out. Many types of entry flaps are available in pet stores and these can be installed in the lower part of the door by anyone handy enough to cut a hole through which it will fit. Make sure it's placed at a height at which the cat can nose her way through the opening—and not have to leap through. Most cats hate to leap as they push through because the door of the flap tends to swing back in their faces. You can hinge a flap to open both ways or one way. This is the training:

1. Tape the flap so it stays open (making sure it won't slam down and scare the daylights out of an

entering cat). Allow the cat to learn where it is and use it for a few days, coming and going at will.

2. Close the flap. Take the cat to the flap's site, put a treat just on the other side so the cat can see and smell it, and hold the flap partly open so the cat has to nose it a little bit at least to open it wide enough for ingress or egress. Do this for several days, encouraging the cat to nose the flap open more and more each day.

3. The last step is to leave the cat on one side of the totally closed flap while you go to the other side and call the cat's name until he pushes through the flap, all by himself, to the waiting reward. Do this for a few days until the cat knows he has total control of his own comings and goings—through that simple little covering that opens his way to freedom—or home, whichever the case may be.

How to Train Your Cat to Love His Carrier

There will be times your cat will have to be taken somewhere, unless you're both hermits, and it's to your advantage to accustom him to his carrying case—even if you can't exactly get him to *love* it. (Some cats *do* love it, as you'll see in a moment.) I have so many clients who complain to me about how frustrating it is to get the cat into the box—often it's a track-and-stalk mission with the person tracking and stalking the cat who has hidden in his most unreachable spot.

I advise taking advantage of the "secret place" affinity that most cats have; and cat carriers, if you look at them properly, are just that: small, dark, private, warm . . . *perfect* secret places.

To begin with, buy a carrier that's permanent—not one of those disposable numbers which become smelly and dank and quite unpleasant because they're just made of cardboard. A wicker carrier is lovely and lightweight but not so simple to clean and disinfect. Other carriers are fiber glass, which are lightweight, and metal ones, which are sturdy. If you line the carrier with disposable baby diapers, which are plump and plush and comforting to the cat, the carrier turns out to be a desirable place to be. But whatever you choose, the carrier should have ventilation holes—many of them. It's not necessary for your cat to have a carrier with a window to the outside; you're not doing him a favor by letting him see the bustle and the activity of the world that's disconcertedly moving past him. That's only confusing and scary. Your cat likes life safe and secure and private.

Get your cat used to the carrier before you put him in it to take him anywhere. Line it with warm, cuddly washable blankets or disposable diapers. The box will come to mean his security place, his nap room, his hiding spot. That's terrific. Give him treats *in* the carrier; create a delightful association with the place. Every cat needs a secret garden and the cat carrier (minus the shrubbery) can be just that. Then, when you need to transport him to the vet or to the car for any voyage, there's no trauma involved in the trip.

Just last week, I was examining Lucy in my treatment room. Her owner had brought her to me for a short stay and had deposited her, in the carrier, and rushed off to an appointment.

My assistant came in with a puzzled look on her face and said, "There's a strange cat walking around the examining room. We have no idea how it got there." We asked in the waiting room, and no one could solve the mystery so I put the visitor in a holding box with a big sign reading: DO YOU KNOW WHOSE CAT THIS IS?

Two days passed, Lucy's owner came to pick her up, spotted the mysterious cat and let out a yelp of joy. "*There* you are, Lawrence!" she chortled in relief. It was her other cat, the one she'd left home (she thought) when she put Lucy in the carrier to come to me. Lawrence Cat, loving the carrier, had sneaked in to take a nap and wasn't noticed when Lucy eagerly trotted in there for her trip. On arrival at my office, Lawrence was again unnoticed when my assistant opened the door for Lucy's exit; Lawrence followed soon after, awakened from his joyride.

"I *knew* that carrier felt heavy," muttered my client, as she lugged both cats home, again happily ensconced in their secret place.

How to Train Your Cat to Sit Still for a Bath

You've always heard that cats are so naturally clean, you never have to bathe them, right? Wrong! Naturally clean, yes, but sometimes (more often than the non-cat owner would suspect), they get unexpectedly dirty and almost cry out for help. A cat that's ill with diarrhea or bladder problems will need to be bathed. A cat who has dandruff caused by a poor diet will need not only to be bathed but will also need to have dietary changes. A cat with an especially oily coat will need an occasional bath. A cat who has had an encounter with a skunk will need a bath (tomato juice baths are the best instant deodorizer, after which a *regular* bath is in order). A cat who's just been through a successful flea treatment needs to get rid of the treatment residue. A cat who's met up with some basically dirty dirt will need a bath. Here's the procedure:

1. First, groom the cat to get out the loose hair and mats.

2. For the very first bath ever, don't use soap or

anything that will take a long time. Your purpose is simply to accustom that cat (preferably kitten) to being wet all over.

3. Place a towel or a rubber mat in the sink so the cat won't slip; she will have a modicum of control over the whole procedure if she is not sliding around the enamel bowl.

4. Use no more than three or four inches of warm water.

5. Place a small cotton ball into each ear to protect from water.

6. Spread a drop of Vaseline around each eye to keep water from entering the eye.

7. Warm a mixture of shampoo and water in a squeeze bottle by letting it sit in the warm bath water. Use baby "tearless" shampoo (or a good, commercial cat shampoo).

8. Hold the cat gently, but firmly. Nothing could be worse than to have him escape your grasp and flounder around, coughing and choking in the soapy water. *Arrrgh.* Keep talking. Lie if necessary, and tell him *you* like baths better than chocolate. Talk to him and tell him that baths are the most divine pleasures. Nuzzle him with your face. Get his paws wet first to give him the idea of wetness. He'll probably like it, but if he doesn't don't take no for an answer. You might trail a piece of cord or a rubber mouse in the water and let him (as you're holding him) paw at it; this is no different from the toy boats and bath toys a baby loves.

9. Gently lather him up with your hand, making love circles on his body—loving, *slow* circles . . . not frenzied scrubbing. As you lather his muscular flanks, his graceful back—tell him how much you admire his body.

10. Now, with a hose whose pre-tried spray is gentle

and steady, rinse him off—but don't let him see the water spray directed at him because it will inevitably remind him of that water gun instruction that says no. You want a bath to be a yes experience; therefore, hold the rinse hose at its tip so the water seems to come softly from your hands. If your cat really would prefer no hose (some cats never get the hang of the spray), try a big sponge. You can finish with your own cream rinse gently dripped all over your long-haired cat.

11. Drying time should be a positively luxurious experience. A dryer set on the lowest setting is nice for a cat who's snuggled in a warmed-up towel on your lap. Move the dryer around so it doesn't overheat any one part of the body and never point the dryer at the cat's face and ears—would *you* like a gust of warm air in your nose? If your cat refuses a dryer session (some cats never get used to it), try a couple of unshaded gooseneck lamps with 100-watt bulbs pointed at a place where the cat is sure to "rest up" from the bathing experience. Cats love lamp warmth as much as they love sunlight and they'll lick themselves insensible as they lounge in the glow of a warm bulb. Note: The bulbs should be out of cat reach.

Tip: Make sure your cat's first bathing experiences last just a few minutes—no longer—even if the cat is not thoroughly cleaned in that time. Getting accustomed to having one's ankles drenched takes time.

Occasionally male cats have a greasy, dark secretion running down one-third the length (starting from the base) of their tails. (Female cats get it also, but not as often.) This is known as "stud tail." The treatment for this is applying cornstarch and working it through the tail—don't bathe the cat. The starch absorbs the secretions. After a few moments, comb out powder and grease. If a cat's tail becomes filthy for reasons other than stud tail, you don't have to bathe his entire body. Simply place him on the edge of a sink (that's been covered with a towel) and dip his tail into a dishpan filled with warm lather. Rub the tail gently, pour warm rinse water on it and groom. *Note:* For a really bad patch of scruff, rub a bit of Vaseline into the patch before you wash it; that loosens the dirt.

Once you get through the bath experience firmly and effectively two or three times, your cat will come to sit graciously through the whole procedure.

How to Train Your Cat to Eliminate in Absolute Privacy

Even the most confirmed dog owner would love to figure out a way he can avoid the daily walks that must take place in rain, snow, sleet—whatever. The nice thing about cats is that they're usually quite immaculate about disposing of their waste products and even a two-room apartment can house a cat without odor. Still, there are some people who hate the sight of the litter box—out for all to see, even the boss and his wife. So they've created a private bathroom for their cats, and litter boxes are invisible to all. Many cat owners love the idea of a cabinet big enough to hold the box built *around* the litter box—and a deodorizer. Doors in the front can be opened to clean out the litter, but on the whole they stay closed to keep the cat's bathroom out of sight. On the side of the box (make sure it's tall enough) is a small opening or flap (see page 108 to learn how to teach a cat to go in and out of a flap door) through which the cat can enter and leave as she wishes. All you have to do is show the cat the opening, scratch her paws in the substrate once or twice, repeat a few times, and your cat has an out-of-sight bathroom. This is perfect for apartments with one small bathroom (or even two), where everyone always seems to accidentally step into or knock over the cat's litter.

Cat Workouts (If Moving Is Healthy for Jane Fonda, It's Healthy for Your Cat)

If some of you, like me, can't quite see the intrinsic value in getting a cat to, say, roll over—think of it this way: physical workouts are beneficial; they encourage circulation, bowel movements, lustrous coats and good appetites. Do not consider "circus tricks" *that require movement* as foolish; they have their place in the general well-being of your cat and if you've the patience to teach these workout movements, your cat, especially your sedentary cat, will only benefit.

It's far easier in any kind of training to get a cat to do something than to stop doing something. As I pointed out in chapter 3, where we talked about redirecting unpleasant behavior, the way you teach the cat to do things is by positive reaffirmation—not negative disapproval. "Sit," "stay" and "come" are direct commands in which you induce positive behavior by rewarding a task well done. Cat workouts are usually taught by catching the cat in the act of doing something you think is cute or beneficial, reaffirming its value by letting the cat know just how wonderful he made your day by that jump, and then—getting the cat to repeat the act for a great reward. These are the acts that make your cat a media event— fun to watch and fun to work with, as well as being a boon to his health.

Jump up: It's fun to see a cat jump on a perch, when told, and it can also be useful if you wish to inspect its

coat, eyes, and so on, for any reason—not to mention its
value as a workout. Start with a kitchen chair or stool.
Wait until the cat's hungry. Hold a treat within the cat's
vision as you pat the surface of the chair. Naturally, the
cat will ignore you. Keep doing it. If the cat will not
respond to the patting, lift and place her on the chair;
then compliment her wildly and give her the treat.
Repeat. After a few training sessions, the cat will jump
on the stool whenever it's patted. Always use the same
verbal command as you train and pat the stool's surface—
something like "Jump up!" When you have your cat
responding well to the pat *and* the command, issue just
the command without the patting. If the cat refuses to
jump up, ignore her and walk away. In a few moments,
try again. Eventually she'll respond to the verbal command, and what's more, you can work it so she'll jump
onto any surface you designate simply at the "Jump up!"
command when she knows a treat's coming. It's impera-
tive that you never give her the treat without the jump.
This is a wonderful thing for a cat to learn because she'll
always wait to be *invited* up to a couch for petting when
you or a willing guest want to indulge in a little mutual
shnoogling (see page 35 for definition of "shnoogling").
Cats that leap on cat-haters are always a problem and a
cat that's used to being *asked* to jump before she does it
is always a delight.

Sit up: Once you've gotten your cat to leap on a
perch and she's absolutely nuts about the trick, try
holding the treat just over and above her head and say
"*Sit up*" before you give it to her. She will instinctively
sit up on her haunches (or should, shortly after she gets
the idea that you are pleased with sit-ups). Give her the
treat. It does not take long at all before your cat will

jump up at the command and then *sit up* at the further command. Make sure you don't allow the cat to grab or paw the treat out of your hands. She's to sit there until *you* actively touch the treat to her mouth. If she absolutely refuses to sit, gently raise her to a sitting-on-her-haunches position and then give the treat. Eventually, she'll do it by herself.

Ring for out: If your cat is an outdoor cat and you hate its yowling and crying to go out—*and* you don't have a cat door, train the cat to ring a cat-height bell hung near the door. Put the cat near the bell and, lifting the cat's paw, paw the bell. As soon as the bell rings, let the cat out. Do this for several days—help the cat paw the bell, and then release her outside. When the cat does it by herself—make sure you immediately put her outside. This is one trick in which food is not given for learning: you don't want the cat to ring the bell incessantly for a food reward.

Over: A workout trick that many cats seem to love is jumping over a stick. It combines getting a treat with exercise with a game—all three of which are delightful to the feline consciousness. What's more, it's fun for you as well and absolutely dynamic when you're showing off to friends. As kids, we all used to play the game: a friend lifts a stick, higher and higher from the ground—and you have to first step and then leap over it without knocking it down. Cats can leap surprisingly high, and with consummate grace.

First, get a stick that you'll use for every session. Place it on the floor when your cat is feeling hungry and playful. With a treat held directly *out* of the cat's range, entice it to step over the stick several times to get it. Do this for three or four days until the cat is neatly stepping over the stick to retrieve the treat. If she tries to walk around the stick, she gets no treat. Each time she's about to cross over, say "Over!" until she associates the motion with the word. Gradually, lift the stick *off* the ground— very, very slightly, increasing the height infinitesimally every few days until she's gracefully stepping over a stick held an inch or so from the ground. You'll have to be patient when the height reaches a level the cat must *jump* to clear. Just encourage her verbally; don't allow her to walk around the stick (teaching this in a narrow hallway is a good idea) and give her time. Patience will eventually breed success. If she absolutely refuses the higher level, go back to the walk-over level for a few days—then slowly raise it again. As long as she physically *can* jump, she will, sooner or later. Of course, after that first jump, be extravagant in your praise. Keep at the trick to prevent trick amnesia—or else you'll have to start from scratch (so to speak) again.

Note: This is the same trick as "jumping through a hoop" (you might have figured that out yourself).

Fetch: A cat I know named Pushkin prides himself on retrieving tiny, rolled-up balls of tinfoil—a game that never seems to tire the person he keeps. Here's how Pushkin learned to fetch:

His owner tied some invisible fishing line to the rolled-up tinfoil and threw it. Naturally, Pushkin took off after the ball as any self-respecting cat will do. When he retrieved it, Pushkin's person slowly pulled both cat and ball back to him. After taking the ball from Pushkin's mouth, the cat was rewarded by a treat.

Five or six times should do it because cats naturally love to bring presents of prey back to their beloved people . . . and if they're rewarded for doing so—fantastic! After a while, remove the line from the silver foil—your cat will think it's superfluous.

How to Train Your Cat to Resist Danger (By Training Yourself As Well)

Of all the commands your cat can learn, "Come," "Stop," "Stay," "Jump"—probably the hardest to teach is "You simply cannot do that any more!" Cats, despite the common myth, are not immune to danger. Much has been written about the curious cat with nine lives. Well, the curious part is correct but the nine lives part isn't, and too many swell cats I've known have ended up with half a life or less because their owners thought they were invincible.

Cats love to explore interesting-looking places. Some of the more interesting-looking places have names like The Oven. The Refrigerator. The Clothes Dryer. The Open Window. Drawn to good smells and cuddly-looking enclosures, cats too often come to disastrous ends because the coddled atmosphere in which they live, loved and loving, precludes an owner from being harshly strict about possible danger. That's a shame. How many mothers who find it anathema to spank their children find it absolutely correct and natural to give their kids a resounding, lesson-teaching spanking when the child is in mortal danger . . . about to run in front of a truck? Now, I'm not saying you should ever strike a cat (a useless punishment), but if you have to teach it a lesson by being what might seem rather cruel—do it! It's important to scare that cat away from what might be the death of it. That's training of a very urgent sort.

For example: Belinda, my client's cat, was entranced by the refrigerator. My client was terrified that one day she'd leave the fridge open, turn away for a moment as the cat, unbeknownst to her, scurried in and then ended up frozen more solid than a cube as the door closed for the rest of the day. She waited until the cat did just that—jumped in to inspect the open cans of cat food. My client, heart in mouth, *closed the door of the fridge,* and for a full five minutes ignored the cat's frantic yowling and scratching to escape. They were the longest five minutes of her life. When she opened the door, standing well to the side, I might add, the cat *leaped out* in absolute terror and scurried, cold and daunted, to her sleeping box. She never, ever went near the refrigerator again. Sometimes, a cat owner has to take a chance. How eong can you leave a cat in a refrigerator before you have frozen feline? She didn't quite know but figured as long as the cat was screaming, it was not yet freezing up.

My own cat had a propensity for climbing into a warmed-over oven, inadvertently left open for a moment. Also not terrific because not only could she be locked in all day by accident—she might catch on fire from the pilot light or, blowing it out, asphyxiate from gas fumes. One day when she jumped into the oven, I did the next best thing to lighting it, which would have been rather too drastic a remedy. I closed the oven door and then proceeded to bang the door with pots and pans—created a *wild* rumpus—until that cat was convinced that the oven was not the Caribbean. She never went in there again. One thing about cats—they're fast studies when they're scared.

And then there was the cat at the Harvard Club named Bootsie Mittens. Bootsie discovered an opening in the ceiling of the Grill Room and it took the combined skill of the entire Engineering Department to rescue him, on three separate occasions. The men of Crimson became

quite attached to Bootsie and allowed him to be inter-
viewed by the press; when his admiring public wrote
him letters, Bootsie answered and signed with a paw
print. The Harvard Club, ever vigilant on matters of
propriety, though, reports that, "An attempt to have
[Bootsie's] biography published was discouraged by the
Board of Managers in line with the Club's policy on
publicity."

Bootsie was eventually discouraged by simply closing
off the opening.

Raining Cats and Dogs

This is not just a cute chapter heading. Every year,
thousands of cats fall from open windows because their
owners believe that they're made of rubber. What's more,
young children who have been told that cats invariably
land on their feet when they fall have been known to
toss Felix right out of the window—confident in Felix's
ability to land upward and safely. Well, cats do have a
wonderful way of righting themselves midair because of
their more than five hundred graceful voluntary muscles
and acute inner ear mechanisms, but they don't always
land on their feet and they're often killed or injured in
falls. This happens so frequently it even has a special
name—High-Rise Syndrome. It is most frequently seen
in big cities and in warm climates, and there's an increase
in incidents during the summer and fall. Some clients
report that a cat just rolled off a terrace when he was
napping. Nonsense. The owner couldn't believe that his
cat would jump—because it never did before. Cats *do*
jump *and* fall from high-rises.

If you live high enough so that a cat can die from a
fall or jump, you *must* screen windows or balconies or
keep them closed. I had a patient who sat safely on the

window ledge of a nine-story building for *years.* One day (my client saw the whole thing happen), a wondrous butterfly drifted by. The cat was much taken with it. It jumped out of the window, forgetting everything but chasing that fairy-tale creature. It did *not* have nine lives. They never do. There *is* no training to keep your cat safe at heights. You must train yourself to always close or shield the windows in your home.

For your information (and because this is the *intelligent* cat owner's guide), the record distance cats have fallen, and lived to tell the tale (according to the magazine *Off-Lead*), is eighteen stories onto a hard surface, twenty stories onto shrubbery, and twenty-eight stories onto a canopy or awning. Don't try to see if your cat can compete: it won't be grateful to you.

Other No-no's

While you're training yourself to keep windows closed, give yourself a refresher course on not leaving poisonous plants, soaps, mothballs and the like around the house. Cats, in their play, can easily chew on a substance that is lethal, and a mothball, for instance, is the perfect example of a chasable, attractive killer. Even the fumes from mothballs can make your cat very ill. Ditto, sprays for insects. And, train yourself to check out every opening you're about to close to see that a mischievous cat hasn't entered to be entombed for as long as it takes to find him when you discover him missing. I know a cat that survived a quarter cycle of a clothes dryer before its owner frantically retrieved it, but the cat was pretty piqued at the experience.

Tricks You Won't Learn Here

You may be able to teach your cat to roll over or meow on command, but you won't learn how to do it in this book. I think rolling over is for the dogs and not for intelligent cats or intelligent cat owners. Rolling over has always seemed a bit demeaning to me, and cats don't appreciate being patronized. As far as meowing on command is concerned, I discourage this trick heartily. There's no better way to encourage the Crybaby personality (see page 67) and if your cat learns he can get a treat from crying loudly, he'll try it at inconvenient hours, say—4:00 A.M.

Cats and Babies Together

Another thing I won't attempt to teach you is how to protect your newborn infant from your cat. Most of the time, it won't need protecting. A cat that's been slowly and lovingly introduced to a new baby will usually be very gentle and loving to it—but, *you never know*. An atavistic, territorial-protection instinct may surface just as your infant flings its arm over the spot in the cradle that the cat has claimed for her own—and that heretofore loving cat may just scratch out to protect its territory. And, although there's no truth at all to the old wives' tale that a cat will suck a baby's breath, a helpless newborn is no match for a fat cat who has decided the baby's face is a nice warm place on which to snooze. So, I strongly suggest that you never leave a new baby and a cat together unsupervised. If the baby has his own room, replace a door (you want to keep open so you can hear him) with a screen, which will keep the cat out and still allow you to hear distress sounds. You are responsible for supervision of the two in the baby's earliest months and no cat training should make you overconfident that the cat will not, accidentally or on purpose, harm that infant. Eventually, as the two grow together, you can surely relax your vigil; cats and babies are wonderful pals, after a while, and love to snuggle together. In the beginning, though, caution is the watchword.

Even when you consider your cat wonderfully trained, practice with him regularly and consistently. After some time passes, and you haven't put him through his paces, it is inevitable that his response to command will weaken. He simply forgets, that's all, just like you forgot a whole lot of things you used to know. Refresher sessions are PAR for the course throughout the life of the cat.

If you wish a well-trained cat, you must accept the fact that if the cat has to perform for you—you have certain responsibilities toward it. Just as you can't "own" a person, I really don't think you can "own" a cat in the finest sense of the word. Also, your cat is a distinct creature with rights and personality, and living comfortably with it implies a dual kind of giving. Dogs have been known to slavishly abide by their owners' wishes—even when those owners concede no rights at all to their pets, but show me a well-trained cat whose owner mistreats her; I do not believe such a cat exists. Sweetly manageable cats are those whose owners understand that cats have wills that must be acknowledged.

Our founding fathers understood about rights and responsibilities and in their wisdom developed a near-perfect Constitution and an even nearer perfect Bill of Rights.

In deference to them, I have prepared what seems to be a reasonable Bill of Rights for your cat and one, as well, for you. Consider them together to be a kind of cohabitation contract, if you will. . . .

A Cat's Bill of Rights

1. My person shall make no attempt to restrict my freedom of movement (except that restriction which can be soundly proven to benefit my physical welfare). This especially applies to holding me when I do not wish to be so held.
2. My person shall not attempt to train me to do ridiculous dog tricks like play dead and roll over.
3. My person shall not laugh out loud at me (or encourage others to do so) no matter how silly I look in the course of my training.
4. I shall have the right to bare claws (if I still have them) to protect myself and my property.
5. I shall have the right to object strenuously if any other cat is quartered in *my* quarters without my consent, during war or peace.
6. I have the right to be secure in my special spot, litter, toys and general catness and to protect same against unreasonable search and seizure.
7. I shall not be removed from my place in the sun without due cause.
8. Under no circumstances whatever am I to be punished by hitting, slapping or kicking or else I shall again have the right to bare claws—or teeth.
9. Dire need must be proven before I am to be immersed in water.
10. My person shall admonish his acquaintances, spouse and children to avoid shutting my tail in appliances or sitting on me.
11. My person shall not leave a halibut to thaw, then get

mad when he returns and finds it decimated.

12. I shall not be squirted with a water or air gun unless I transgress against numbers 3, 9 and 13 of my person's bill of rights.

A Person's Bill Of Rights

Fairness dictates that the cat with whom I live, may NOT

1. Hang out with skunks or in briar patches
2. Loll around in grease, chewed gum or dirty rainwater
3. Eat the geraniums or the phlox
4. Spray foul-smelling evidence of his lust around my dinner table
5. Nag, nag, nag
6. Trip me when I am in a hurry
7. Vomit in my shoes or in the shoes of my beloved
8. Deposit mites or fleas all over my person
9. Willfully excrete in places not specifically designated for excretion
10. Bring me his mouse prey and expect me to freeze it for later (a position the cat-who-lives-with-the-editor-of-this-book unashamedly takes)
11. Require that I feed him the best cuts from my plate
12. Stomp loudly all over the house in the early morning hours
13. Destroy the Louis Quinze chair when he is piqued at me

5

Cat Communiqués

You Can't Train Your Cat if You Can't Talk to Him

The Big Messages

Your cat has something to say about training as well as about how you will coexist with him. Cat language will be understandable to you if you pay attention, and your language, in great part, is understandable to the cat— both in content of certain key words and tone of voice. I firmly believe that we who own and love cats *must* consider cat communiqués not as sounds from an alien species but as the endearing and comprehensible messages of close friends.

Your cat smiles, freezes, shows love, terror, anger in the way she holds her tail, head, body, hair. People who spend time discussing matters of importance with cats (as every cat owner should from time to time) come to know, as surely as though the cat were speaking *words,* just what she wants, thinks, needs. Although every cat is unique, has her own adorable way of saying, "Let's play," "I'm bored silly" or just "Stop that immediately!" certain cat communiqués are universal. There is no question that you must learn to hear what your cat's saying before you attempt training. Just think: suppose you were trying to toilet train your toddler. Every time you put him near the toilet, his face scrunched up in terror, his body froze rigidly and he cried. You'd get the clue that something about the toilet was turning him off, wouldn't you? You'd know, by reading his body language and hearing his sounds, that before you could teach him anything, you'd have to make him more comfortable with the bathroom, the sound of the flush, the gaping hole in the toilet bowl. His sounds and his body would tell you he wasn't ready for a toilet-training session.

In this respect, cats are no different from people. You must surely be able to read the communiqués your cat is sending out before you and he can accomplish any meaningful learning.

The only way to do this is to get into the habit of talking with your cat and carefully paying attention to her responses in terms of her body language, purring, her facial expressions—the whole gamut of cat communiqués. The important thing to remember is that you have to set up a *routine* conversation with your cat to have conversation be meaningful. Cats love the familiar and if you are in the habit of confiding in your feline, telling her about your day, asking her advice about your business matters—your cat will respond by being a communicative creature. This may sound arch, silly to the intelligent cat owner at first, but it has everything to do with having a wonderful cat. If you treat your cat like, say, a *cat*—without giving it the dignity and respect of your conversation, you will very soon have a taciturn, sulky, people-avoiding cat. If you communicate with your cat as if he were a sentient creature (which he surely is), one with the capacity to understand, love, hear you and respond—you will have the king of cats (or queen, as the case may be), the friend of friends in your life. Lewis Carroll has Alice saying, in *Through the Looking-Glass,* that kittens always purr. "If they would only purr for yes and mew for no, or any rule of that sort," she says, "so that one could keep up a conversation! But how can you talk with a person if they always say the same thing?" Wrong, Lewis. They *do not* always say the same thing—you just haven't been listening! It's clear that Mr. Carroll had never mused, chatted, shot the breeze with a kitten or a cat or he'd never have said anything so nonsensical.

Cat training, in huge part, has so much to do with understanding your cat's moods and needs through mutual communiqués. If you and your cat have the wit and intelligence to set up a strong conversational pattern, you and she both feel less silly about crossing the conversational barriers of human/animal conversation every day you do it.

What follows is sort of a dictionary of feline body language and sounds. It is quite accurate—up to a point, and that point is your own cat's individual language, which *may* differ (a cat whose tail suddenly becomes busy is *never* feeling peaceful) but, on the whole, certain communiqués are standard in almost every cat. Smiles are universal in humans, for example, no matter what color, religion, age you are—with *few* exceptions. And, with cats, for example, a humped, arched back is standard "felinese" for "I'm *furious.*" Here then, is how your cat sends the big messages to you—tells you what bugs him, what pleases him, and whether or not he's in the mood for love—or learning.

Body Language

Tongue Talk

The tongue lick speaks volumes. Remarkably versatile with their tongue licks, cats say hugely different things with them, depending on the situation.

Embarrassment: If your cat is trying to change the subject because it feels foolish or guilty, or if it is chagrined because you've firmly said the "no" word or have caught it in the geraniums, it may begin to lick itself—nonstop. This frantic kind of self-lick, this urgent and *committed* tongue action is very different from the long, slow, musing kinds of licks the cat employs to groom itself, for example. The embarrassment lick usually sees the ears splayed back and the eyes half closed. This is not an optimum training time. Your cat has to get over the chagrin at being caught stomping around on the dining room table.

Affection: Every now and then, if you are extremely lucky, your cat will lick you with affection. *If* you've been good. You'll know this lick when you experience it. It feels like the touch of a good friend. Delicious. *Good* time to train.

Permission seeking: A cat will lick you or another cat if it's seeking permission to climb on your lap or on the couch another cat happens to be occupying. Short, staccato licks with the cat looking at you and poised for flight (if the answer should be no) are usually the pattern. The cat's feeling alert and possibly ready for work.

Boredom: A cat that is really bored—bored to the point of desperation—will constantly lick herself in deep, intense licks, sometimes causing skin problems and hair loss. If your cat is licking herself far more than you believe is necessary, if her coat or skin is looking a little ragged—try a little affection or a new cat or a few new toys to take her attention away from the obsessive grooming. *Good* time to work your cat.

Nerves: A cat that's nervous or has an excess amount of energy will lick itself inordinately in short, shallow licks. This cat also needs diversion and extra playtime to work off its tension. Training is difficult when your cat's tense.

Neatness counts: A cat really believes this. The *most* common lick is the grooming lick. The cat's rough tongue allows for a wonderfully efficient, loofahlike sponging. The rough tongue can cause problems with overgrooming, though, because irritation and skin inflammation can result. Certain cats (notably Siamese and Abyssinian) tend to lick and even bite one or more areas more than other body areas. If your cat is grooming herself raw, certain veterinarian-administered medications such as corticosteroids have been found to cut down on excessive grooming. Not a bad time to work on a few commands.

Okay, okay—let's have a truce: If you've ever seen two cats fighting (or hissing and threatening a fight) and one of the cats suddenly sits down and in an amazing show of cleanliness becomes absolutely preoccupied with grooming itself, that cat is saying, "Enough already—I kind of give up and I'm licking myself to save face."

Contentment. When a cat's licking herself and she's feeling warm, relaxed and happy, grab the opportunity to work on whatever training you'd like to accomplish. However, if you see your cat licking herself because she's embarrassed, anxious or nervous—and you can definitely tell the difference after a while—the time's *not* ripe for working on commands.

Head Talk

The cat that makes love to you with her head nuzzling and rubbing against your hands or body is the cat who creeps into the heart of her owner. Head nuzzling always means "I love you and I feel like working."

Tail Talk

On the other end of the cat is a very model of expression—the tail.

When it's flying high: All's terrific. Your cat's not fearful, not particularly hungry, not angry. He may be very excited because you or a yeast tablet is coming. He is feeling *good,* man. He's ripe for lessons.

When it's half-mast: All's not *so* terrific. She's a little afraid or maybe unhappy with the food you've given. Perhaps she's spotted the carrier and doesn't feel like taking a trip. Leave her alone.

When it's drooped low: This cat is very unhappy—perhaps frightened, perhaps even in retreat. Something not very nice has happened to discourage her mightily. Leave her alone.

When it actively twitches or lashes back and forth:

Back off! Now's not the time for discussion. Don't even approach the cat to offer affection—it really wants to be alone. Sometimes a cat will twitch her tail when another cat has usurped her favorite nap spot or dropped something unpleasant in her very own box of litter.

When just the tip twitches: I've always felt this to be an example of extreme self-consciousness, as in when the cat knows you're talking about her and doesn't love all that gossip. Check your cat out when she's taking a snooze and quietly mention her name several times to someone else—and watch that tip twitch. It's funny.

When it's bushy: Anger! The kind of anger associated with attack and maybe the threat of another aggressive animal is often shown by a fluffing up of the base of the tail, and much of the coat besides. This is designed to make the cat look larger than life and scare off the invader. A startled cat will invariably ruffle up its tail automatically. A bushy tail often goes end-in-end with a humped back and a good deal of hissing and spitting.

When it dips between the legs or is wrapped around them: Fear. This cat is quite frightened of something!

Ear Talk

Ears alert and straight up: He's ready for fun or training or affection—the point is: he's *ready* and in a good mood.

Ears flat out sideways: "What's up?" the cat is saying. "What's that noise?" "What's going on around here—

where's my special box for hiding—I'm a little bit nervous around here, guys."

Ears pulled downward: "I'm feeling defensive . . . keep your distance."

Ears pulled downward and to the rear: "I'm really furious! Watch out!"

Eye Talk

Eyes wide open and looking straight at you: You've got his attention. He's interested in what you say and do.

Eyes half closed: He's either sleepy *or* slightly wary.

Eye pupils in slits: He's alert and quite confident—maybe feeling territorial.

Eye pupils "bug-eyed" and round: He's frightened—be careful.

Eyes blinking and winking: He's talking—sharing his day—expressing his devotion—nagging for a catnip fix.

Eyes clouded: Cats have a murky, grayish membrane that ordinarily remains stationary within the bottom of the eye, invisible to others. When the cat is ill, sometimes this membrane is visible. Occasionally, it moves over the eye when the cat is relaxed and being petted.

Eyes staring: This says, "Keep a proper distance." A stare is a challenge. When a cat crouches *without* staring at you, it's feeling submissive.

Paw Talk

Paw kneading: Watch a kitten as it nurses—it kneads its paws against the soft body of its mother, very much as a baby will stroke and touch its mother's skin. Paw kneading means contentment.

Paw nagging: An impatient cat will often give its master a clout or a pull from one paw—this means, "Hurry up, give me a snack, give me a toy, a kiss, or *something.*"

Paw hugs: Cats will often hug with their paws—actually exert pressure in a kind of caress: this is the most endearing of acts and melts even a confirmed cat hater.

Paw blows: Paws and claws are traditionally a defense weapon and a cat attacked will respond by lashing out with paws and unsheathed claws. One doesn't have to be adept at reading body language to recognize a paw blow.

Posture Talk

The cat's body posture is positively eloquent.

Head up, back level and straight: Relaxed and alert!

Sideways stance: A frightened or timid cat will present himself sideways, arch his body fur, arch his back and hope for the best.

Frontal stance: An angry, threatened but very brave cat will face a foe straight on, with an arched back and a hiss—perhaps even show some teeth. Beware! The cat will attack you, unlike the more timid cat in the sideways stance who will just wait and hope you go away.

Lowered crouch: This might be a way of saying, "Okay, I give up . . ." *Or* it might be preliminary to the cat's rolling over and clawing out angrily—either way, it says to be cautious while handling.

Lying on his back: A cat that's lying, looking very contented, on his back *may* be begging you to scratch his stomach. On the other hand, he may be *daring* you to do that and when you do, he will strike out. A cat lying on his back often is instinctively protecting his vulnerable spots (his spine), and he can't help clawing— a kind of mechanical reaction. Be careful.

The Halloween cat: This little black guy with his back arched and his hindquarters elevated in a slight crouch is giving double messages to the world: he's in *both* a defensive and offensive position. His crouch says *defense* and his elevated hindquarters say *offense.*

Nagging

If often happens that a cat develops an annoying and potentially destructive habit and the owner, without knowing it, feeds it fuel to continue. An example: A client of mine had a cat who constantly licked one spot at the base of its tail. The area was practically hairless from attrition. As soon as the cat began to lick the spot, the client (it was finally determined after discussing the matter) would race to the cat, pick it up and stroke it in an attempt to divert the cat's attention. What was happening here? Simple. The cat was communicating with his person (my client), nagging her for attention. The clever cat learned that *lick my tail* equals getting picked up. We proved the point by having my client begin to leave the room every time the cat began to lick its tail. The cat followed her, and began to lick again. Never did the cat (and this was determined by great spying activity) lick its tail when my client was not in the room. Cats do communicate through body language . . . they even learn to nag.

Nose Talk

If you've ever had a cat rub its nose across your face, or brush its mouth on yours, you know you've been made love to. A gentle training session is very much in order.

Whisker Talk

If you accidentally (or on purpose) touch a cat's whiskers, it will close its eyes. This means little, I believe, except that the whiskers tend to protect the eyes by announcing when a piece of dust or a blade of grass is on its way. Sometimes a cat, in great affection, will rub its whiskers across your hands or face—an indication of great affection.

Whisker-twitching: Signifies a scouting-out of food or another animal.

Whisker-droops: A bored, listless or ill cat.

Sound Communiqués

Purring

Now we're into sounds. The purr, traditionally, has always been thought of as a sign of great contentment—which, indeed, it may be. It's interesting to note that purring doesn't come from the cat's vocal cords but from (most scientists think) a vibration of blood vessels in the chest area. If you can't hear the purr of the cat, you can always feel it by stroking (very softly) the cat's throat and neck area. A cat's purr, while it surely expresses contentment, can also express nervousness and even pain. Interesting thought: Have you ever heard a sleeping cat purr? No. I don't know why.

Hissing

If you never saw a cat before in your life, you would not mistake a hiss, especially when it's accompanied by an arched back or a bushy tail. *Keep your distance.*

Yowling and Growling

A cat in heat will yowl all night long. If you've never heard yowling before, you'll know it when you hear it. A mother cat will growl deep, throaty, dangerous growls if she thinks her kittens are threatened. She's telling you not to bother her with commands and training sessions when she has kittens to care for, for God's sake!

Meowing

This is an attention-seeking sound. It may be a meow of frustration or need. Meows come in all lengths and intensities.

Chirring

A twirly, rolled meow given by a mother cat when calling her kittens or by any cat calling his person or other cat pal.

Mating Call

This is usually a closed-mouth murmur and signifies a female's readiness for breeding.

Talking Back

And what about your own body language? Cats will definitely pick up messages, right or wrong ones, by the way you look and act toward them. If you have a child and you never look directly into that child's eyes, she'll feel greatly deprived. In the same way, a cat has got to see bright, affectionate eyes, feel a warm and compliant body if it's to grow up to be open, inquisitive and loving. If you have a husband or lover who is uncomfortable around your cat, you'd be well advised to try a little behavior modification on the *person* who spends a good deal of time in your home. Just as a husband (who feels initial jealousy at the time his wife spends with their new baby) learns to become closer to an infant by changing him, waking up at night to soothe him, playing with him, a slightly insecure-around-a-cat friend can have a much happier coexistence with a cat if he/she cares for it, feeds it, changes its litter, is given some cat responsibility. A cat sizes up people very quickly. You're part of his territory, don't forget, and once he marks you—puts his scent on you by rubbing against you—it's a good idea to respect his claim on you by acting warmly in return. Otherwise, a nervous or unsure-of-cats person is liable to be driven crazy by a cat who *wants,* even demands, easy companionship.

A Note: I've found it helpful to reward the *friend* who's insecure around cats with the positive behavior modification method. A woman whose boyfriend hates her cat might find that boyfriend more kindly disposed toward it if she rewards him with a gentle touch, a kiss, a great dessert every time he is gentle with the cat. The girlfriend of one of my clients "accidentally" left the windows open in her boyfriend's house every time she left it . . . a huge danger to the cats who also lived there. My client purposefully showered more attention on her than he did on the cats, and after a while she learned to coexist with the animals, closing windows and all (she never learned to *love* them!).

6

A Potpourri
of Cat Smarts

Your Cat's Psyche

Understanding Behavior: The Surest Route to Training

Some Thoughts

• Cats are not terribly conversant in rules of property. They don't understand about "fairness." If a cat claims a piece of territory, either accept it, or train him away from it, if possible. Don't think the cat's being "mean" if it usurps your (or another cat's) favorite chair. It simply takes what it wants—that's the rule of nature—until convinced otherwise.

• Cats don't believe in murder. If it brings you a gift of a dead bird or mouse, don't punish it or shriek in disgust. Accept the present, get rid of it, and understand that your cat was making you a love offering.

• You *can* develop your cat's personality and intelligence by coaxing its hunting instincts into sharp focus ... even if it's a house cat. Trail a thick rope along and allow the cat to pounce and stalk; throw rolled-up balls of crinkly paper for it to retrieve; throw catnip mice to stimulate the chase. Cats who are never encouraged to "sharpen" their charisma are dullards.

• The best cat toy in the world? A drip or a tiny stream from a faucet. The cat thinks it's something alive that must be caught and he can play for hours in the sink. Just make sure the drain's open to avoid a flood. (Naturally, environmentalists will hate this idea, but look—how much water are you wasting with a half hour or so of drips?) The next best toy in the world is an empty tissue box with a toy mouse or ball inside. Make sure the opening in the box is wide enough for the cat to swipe the ball out—or else he will be mightily frustrated. Remember that cats adore well-hidden things. The *third* best toy is a length of rope with a frazzled end, tied to a doorknob. Great for many swipes during the day.

• A cat that's exhibiting behavior problems as a result of a new stimulus inside the house should be gradually, *very gradually,* exposed to the stimulus for increasing periods of time. The new stimulus could be a new person in the house, a new animal, a new kind of noise: desensitization takes place if the cat is calmly and lovingly talked to and petted as he's getting used to the new "thing."

Here's an important concept: Social stress makes a cat more susceptible to disease, and stress and fear can definitely affect the digestive process.

• A cat's aggressive behavior may well be stimulated by something that's invisible to you. A cat coming home from a kennel with new smells all over her may evoke a response from her best friend at home that's surprisingly hostile. A cat that's frustrated at the feeding bowl may attack your drapes or even you. These are called displacement attacks: before you can hope to train your cat, you must become sensitive to the stresses that may trouble him. For instance, you'd never *start* training a cat to come when called if recently there'd been an addition of a new cat in your household; the old cat might be feeling quite piqued and not up to learning at times like these.

• Want to avoid inter-cat fighting? Raise your cats together as kittens: it's not necessary to have them be littermates. An adult cat will be happier with the addition of a kitten than with another adult. One of a different sex, though adult, has a better chance of being accepted than one of the same sex.

• Two cats are happier than one cat but there's no great boon in keeping three, four or more in a house. In fact, the Barnyard Effect (see page 43) occurs when a cat feels stressed by overcrowding.

• What makes a cat antisocial with its peers? Any tabby whose owner anxiously hovers over him when he engages in roughhouse play with other cats is sure to believe that other cats mean harm to him.

• Golf balls are great toys: you can't destroy them; they roll appealingly; they even roll under things, tempting a cat to retrieve them . . . a honey of a toy!

Your Cat's Doctor . . .
A Cat's Best Friend (Next to You)

Just as you would find it impossible to give a child every health advantage without having a doctor you trust, a cat needs his own special doctor. In every field, there are disinterested, as well as incompetent, practitioners along with the truly wonderful professionals who make life easy because of their calm and competent manner as well as their informed expertise.

I had one client come to me with a horror story that

I find quite typical of impersonal care. She'd brought her cat to a famous animal institution because the cat was coughing. A sketchy history was taken, the cat was subjected to a myriad of X-rays (every part of that cat's body was X-rayed) and she was asked to leave her pet for further workups. When she returned to pick it up three days later per direction, the cat was in disastrous shape, bite marks on her body, matted fur, glazed eyes. Not only was the cough still present, the cat seemed nearly dead from exhaustion. My client was presented a bill of over $700 and an option: leave the cat to be destroyed (nothing could be done for the illness, she was told), or pay up and take the cat home. She paid and retrieved her cat. It took several months of treatment to bring that animal back to itself, and the trauma and fear it must have suffered from ill handling was exhibited for the rest of its life, as it hid from all strangers.

Here's what I suggest are important considerations in choosing a doctor for your cat:

1. Is the clinic clean? The place where the doctor practices, either private office or clinic, should be sweet smelling and clean.

2. Does the doctor show a willingness to listen and does she answer in language you can understand? Anyone that batters you with medical jargon that's incomprehensible, and refuses to explain (or explains but acts as though you were a fool for asking), is *not* the doctor to choose.

3. Does the doctor show a respect for your ideas? Be wary of one who says or acts as if your comments and opinions are worthless; good doctors make connections between their own observations and those of the owner. You are the key link in your pet's health.

4. Has your doctor taken a complete history which she writes down on a chart? Careless veterinarians can't be bothered with the past.

5. Does your doctor explain the need for each test and does it sound reasonable? Sometimes, you have to use your own judgment. The doctor who prescribes an *entire* set of X-rays for a cat in a nonemergency situation each time you bring it in for an opinion is not using skill to determine if the X-rays are really necessary. Too many X-rays can be harmful to animals just as they can be to people.

6. Is your doctor honest? Few specialists know the answer to every problem immediately. He or she may have to research a problem or even advise you to see a specialist in, say, nutrition, if that's not a field he's totally familiar with. Respect honesty—acknowledging limits is not a sign of incompetence.

7. Finally, and this is the crucial point: Does your veterinarian seem to *like* animals? Is she compassionate, warm and *respectful* toward them? I know too many specialists in people *and* animal medicine who hate their work. I think that makes a poor doctor.

Your Cat When You're Away

If Princess Di and Prince Charles invited you for a cruise on the royal yacht, it would be bad form to ask if you could bring the cat. Even the most dedicated of cat fanciers comes to a day in her life when she simply cannot take Lavinia with her, and so she must make arrangements for that cat's safekeeping in her absence. Several possibilities exist, and in order to illustrate the importance of choosing the proper one, I offer this charming story from the book *The Cat's Pajamas* by Leonore Fleischer (Harper & Row, 1982).

I call this story

The Cat's Revenge

I once knew a fellow named Seth Stone, who was one of life's losers. Charming, affable, well-educated, intelligent, Stone possessed a will to succeed that was surpassed only by his will to fail. For every steak tartare he washed down with champagne at the Four Seasons, he spent a night of fasting and prayer at the Salvation Army. In one of his periods of famine, some Europe-bound friends asked him to cat-sit. It meant a roof over his head, and Stone accepted. The friends, a childless couple, owned a Siamese cat that was not only the apple of their eyes but the whole damn orchard.

Having given exquisitely detailed instructions for the care and feeding of Dingaling—fresh chicken livers, sautéed lightly in equal parts of butter and sherry, two large mushroom caps—and having left behind $35 to purchase the livers from the best butcher, the couple set sail for Le Havre.

Seth went out at once and spent the $35 on cigarettes and beer. Now they were penniless.

During the first week, the cat sulked and refused to eat all but a mouthful or two of the canned sardines and tuna Stone found in the pantry, and scratched earth over the scraps from the refrigerator.

During the second week, the cat changed its tune and ate anything, but there was little of that.

By the third week, they were totally out of food, and the cat, desperate, would have eaten paint off the walls.

Stone was able to cadge an occasional meal for himself from a friend, and there was always the promise of a Salvation Army handout, but the cat was in Hard Luck

City. It complained day and night, as only a Siamese can.

On the morning that the owners were to arrive back home, Stone took his first long look at the cat. What a mess! Dull eyes, uncared-for coat, and its ribs sticking out six inches on either side. It looked as though it hadn't had a meal in a week, which it hadn't. It dawned on Stone that he was about to face a firing squad. The minute those people came bursting through the door yelling for kitty, and saw the pathetic broken creature it had become, Stone's life wouldn't be worth a telephone slug. He had to do some fancy dancing, and fast. Racing through the kitchen, he began flinging the cupboard doors open, the cat howling plaintively at his heels. Not one sardine, not even a Ritz cracker. In rising panic, Stone made for the freezer. Empty, except for two packages of frozen green beans. It was time for desperate measures. He ripped open the packages, put the beans up to boil, and poured the cooling mess into the cat's dish. Dingaling fell on the saucer and wolfed down the beans.

As soon as it had eaten, it became a different cat. Its sides filled out; its fur took on a sheen. It was a Siamese again, not a piece of Bowery rubbish.

On cue, the door was flung open and in rushed the couple, calling for their pet, falling on its neck with little whimpers of adoration.

"She looks terrific! Did she give you any trouble?"

"Are you kidding? She and I are the best of friends. Chicken livers every day at ten o'clock, just as you ordered."

That mendacious bastard! If looks could kill, Dingaling would have fried Stone where he stood.

Instead, the cat gave a long shudder, one mighty convulsive heave, and threw up two packages of frozen green beans.

It's important, you see, to give some close thought

to the pedigree of the person or persons who will be responsible for Lavinia in your absence.

Cats love to follow a tried and true routine, so there's no question that given the choice of the Waldorf-Astoria when their owners are away or their own, humble home, the home wins out, every time. There's far less stress involved in having someone dependable come in to feed, clean litter and bestow a few hugs every now and then than boarding the cat in even the finest of catteries. New environments invariably make cats uncomfortable. *So,* if a friend stays with, or looks in on your cat daily, try to make sure the cat and the new person are friends before you even leave. Be careful to leave written instructions as to feeding schedules (most cats become furious if that food's not in that dish at the accustomed hour), medications, special problems, and the number of your veterinarian. Tell the caretaker not to leave windows, fridge or oven doors open. And ask your friend to change the cat's litter regularly because a cat will ignore a dirty litter box and find another convenient toilet—usually in your closet. If your friend dislikes changing the litter, try leaving a package of plastic trays that fit in the box; this way, the whole tray and litter can be removed and a new one substituted without your friend having to touch the dirt. Naturally, if there are two cats to leave, one provides company for the other and that's much better than solitary confinement.

If you have no friend to come and visit, you could think about leaving the cat in the friend's house—but *only* if the friend has no cats or dogs. Strange cats always present the danger of your cat's picking up mites, fleas or infections, and the established cat is going to be less than thrilled with an intruder. Cat fights are a strong possibility—and, who needs it? Dogs, even gentle dogs, may create stress simply by their odor and throw your cat's sensitivity level into wild disorder.

I recommend a professional, clean cattery (not a kennel where dogs are also boarded ... that dog smell, again, makes felines *nervous*) if you haven't a responsible caretaker to come in. Check it out before you leave the cat there; Lavinia cannot pack her bags and leave if the service stinks. A cattery should not smell, should have a veterinarian on call twenty-four hours a day, should have enough staff, proper heat, ventilation and compassion; you can sense *good* vibes—do the people at the cattery look like they *like* cats? Trust your instincts.

Ask to see the cages where your cat will be kept; if the floors of the cages are constructed with wire, don't choose the place! Wire bottoms have a tendency to catch and maul little cat feet. Cats should be thoroughly examined by the doctor before she accepts your pet, and inoculations should be up to date. Cats should not be in contact with other cats—there's no better way to spread problems. Females in heat should be placed in separate quarters because they can drive the whole cattery nuts! Finally, give your cat her favorite blanket or toy to take along on her vacation; it's comforting and a way to while away the hours until you return.

If you don't know a good cattery, ask your veterinarian to recommend one. She might even be able to suggest a professional cat sitter who will come to your home and stay, or just feed and clean the cat when you're away. Bon Voyage! It'll be okay—I promise ... Just keep the cat sitter away from the frozen string beans.

Your Cat (Choosing the Best Candidate for Training)

So, you want to have a cat that's wonderfully trained, do you? Think about that *before* you buy the cat. The newest medical and investigative research says that choosing the right cat has much to do with its potential for training—and that, of course, means genes and the earliest conditioning the cat's received.

Let's talk hereditary characteristics first. It's difficult to train white cats with blue eyes for the simple reason that many of them are born deaf. (Try teaching a deaf cat to come when called.) Now, you may love white cats with blue eyes, and that's just fine; don't, however, expect them to be the most prime candidates for training. (Test for deafness by standing behind the cat and making a loud noise—like popping a balloon. If the cat doesn't react, you know it can't hear.)

We know that genes count in the temperamental characteristics of cats. Domestic cats, for instance, by virtue of their genes are far less nervous, less aggressive, more easily trained and handled than their leopard forefathers. And certain domesticated cats seem more gentle than others—longhairs, for example, are for the most part calmer and more docile than shorthairs ... even though there are many, many exceptions to the rule. Ordinary street cats can be the most wonderful in the world, but since their genetic origin is mixed, it's difficult to predict their personalities and training possibilities. Still, I've known thousands of pussy-cat street cats.

Among the cat purebreds, some are particularly

known for sweetness and tractability (although, again, there are always exceptions) and educating yourself by doing some reading and talking to breeders is always a good idea before you decide to buy a specific breed. Some of the breeds that are more amenable to training are the Himalayan, Persian and Turkish angora longhairs and the Burmese, Russian Blue, American Rex and Havana Brown shorthairs—among many others.

Even more significant than genes, in my opinion, is the conditioning the cat's received from the moment it's opened its eyes.

Kittens are the best bet to bring home because they'll find it easier to adjust to your life-style than a mature cat who may have had poor experiences with people. That is—most kittens. Although I don't like to generalize, the runt of the litter, the very smallest of the brood, is probably likely to have the most problems. Because it's smaller, it has had a tougher time obtaining its mother's nipple during nursing—and behavioral and nutritional problems can easily result. Undernutrition in the earliest stages of life can result in brain damage as well as vitamin and mineral deficiencies which may be permanent. This has a huge bearing on behavior and training, of course. Furthermore, the runt of the litter is frequently picked on by its brothers and sisters and thus has had an early jaundiced view of companions. Many excessively aggressive or excessively shy adult cats have been found to be runts of litters. They are simply not the most relaxed kittens in the world—who would be if they had to fight for survival?

It's wise to choose a kitten that's been born and bred in a place that's clean and relatively quiet, because dirt breeds disease and noise breeds nervousness.

Interesting Tips

• New information from the Cornell University Feline Health Center says kittens that remain with their mother for two to four months rather than four to six weeks (as is the routine now) have improved sociability with other cats.

• Veterinarians have discovered that very tiny kittens, handled *for just a few moments* every day by a human and then returned to the nest with their mother, are wonderful subjects for training. The handled kittens are "less emotional in strange environments and most resistant to some disease."

• Don't worry about breaking the kitten's heart by removing it from its mother. Neither mother or kid much care and all memories of each other fade almost immediately when each is treated lovingly and happily. If they should meet again in a month or so after separation, each might think of the other, "What a nice cat" and go on her way without a look back.

In the end, a well-trained cat has been a well-loved cat and a well-cared-for cat. A cat that's been mistreated from the beginning will never respond well to a human directive. It has no reason to believe the human is out for its good.

I started this book by telling you: "You can so train cats!" It's true. You can. But, I also told you that in so doing, compromises must be made—that although cats are smart and willing and able, they're in a different league from dogs when it comes to temperament. You can train cats to do what you want them to do—most of the time—which means that if one day your perfectly reasonable, loving cat decides, for no reason at all, not to come when he's called, or not to walk on his lead, well, you know how cats are.

Which brings us to a perfectly wonderful article I read in the Boston *Globe* one day, titled appropriately, "You Know How Cats Are." I'd like to end our book on the note that writer Diane White sets:

You Know How Cats Are

Mom finally taught the cat to catch a Frisbee. It took a few years, but Mom says the time was well spent.

You may wonder why anybody would want to teach a cat to catch a Frisbee. We all wondered, too, at first. "Fern," my Aunt Abby said, "why in the world are you teaching that cat to catch a Frisbee?"

"Why not?" Mom said. "Dogs do it."

Mom said she saw a dog on television who has made a career of catching Frisbees. This dog, she says, goes around the country with his owner, all expenses paid, just showing off.

Well, Mom always says cats are just as smart as dogs, smarter maybe. Cats can do anything dogs can do, Mom says, if they're in the mood.

Personally, I suspect Mom had visions of herself and the cat traveling around the country, all expenses paid, showing off. Not that she ever said anything about it. She only said she was trying to prove her point.

So, about three years ago, Mom started trying to teach the cat to catch a Frisbee. It wasn't easy. The hardest part, Mom said, was learning to catch a Frisbee herself.

"You know how cats are," she said. "You can't just tell them to do something. You've got to spark their interest."

Mom thought the best way would be to go out in the back yard and practice catching a Frisbee with her own teeth. "It was tricky at first," she said. "The key is not to get down on your hands and knees, but to stay kind of semicrouched, ready to spring."

Mom needed somebody to toss her the Frisbee, so we all took turns at first, all her relatives and friends. After a while, though, the novelty wore off, and most of us quit. But Mom was still game. She started paying some of the neighborhood kids to throw her the Frisbee while the cat watched.

It was hard on her bridgework, Mom said, and sort of humiliating when one got past her. When she'd miss, the cat would look down his nose at her, so superior. You know how cats are.

Eventually, though, it began to pay off. Mom got really good at it. There was almost nothing she couldn't catch. And after a year or so, the cat began to show some interest.

From that point on, it was easy, Mom said. Within just a couple of years, the cat was jumping three or four feet in the air, catching Frisbees right and left.

But that wasn't good enough for Mom. She wanted to train him to do tricks, like that dog she saw on TV. First she taught him to do a sort of half-gainer. That was kind of hard on Mom. She kept wrenching her back when she was showing him how to do it. Then she taught him a flip, then a double somersault, then a double somersault with a twist.

Pretty soon they had built up a whole repertoire of stunts. It was uncanny. Even Aunt Abby, who had had serious doubts about the whole project from the beginning, had to admit it was really something the way that cat came on.

The big moment arrived a few weeks ago. A photographer from the local paper showed up at the house to take pictures. Mom was excited. She didn't say anything, but we all knew what she was thinking. First the local paper, then the *National Enquirer,* the *Guinness Book of*

World Records, "That's Incredible!" Letterman, maybe even Donahue.

We were all out there in the back yard. The photographer was poised. Mom tossed the Frisbee. The cat just sat there. She tossed it again. Nothing. And again. Nothing. After 20 minutes or so, she gave up.

We all felt bad for her—everybody except the photographer, that is, who kept saying he was the victim of a hoax, when he wasn't making threats. But Mom was philosophical. "I guess he just didn't feel like it," she said. "He's not in the mood. You know how cats are."

Index

Abcesses, 18
Accidental defecation, 14
Acidity of urine, 29, 30–31
Affection:
    cats' need for, 5, 8–9, 67
    expression of, 51, 86
        by licking, 130
    and training, 96
Aggressive behavior:
    causes of, 22–23, 142–143
    training to correct, 35–53
Air gun, 41, 50
    discipline by, 71
Alcohol rub, 20
Allergies, diet for, 31–32
Alternate litter-box materials, 60
Aluminum foil:
    to discourage spraying, 57, 58
    to protect house plants, 77
Anal glands:
    blocked, 14, 24
    removal of, 15
Anemia, 18–21
    diet for, 27–28
Anger, and behavior problems, 56
Angry cat, body posture of, 135–136
Antidiarrheal medications, 14
Antisocial behavior, 85–86, 143
    causes of, 38–53
Appetite loss, 18

Arched back, 39, 129
Arthritis, 15, 16, 24
Ash content of diet, 30
Aspirin, 19, 20
Assertiveness in training, 90
Asthma, 19
Attention:
    and behavior problems, 55–56
    crying for, 70
    need for, 67
Author, veterinary training of, 4–5
Availability of food, 84
Avoidance of danger, 119–122

Babies, and cats, 123–124
Back:
    arched, 39, 129
    lying on, 136
Bad breath, 18
Balloons:
    to protect furniture, 78
    to protect house plants, 77
Barnyard effect, 43, 143
Bathing of cat, 111–114
Bathroom:
    litter box in, 62
    private, for cats, 114–115
Bathtub:
    elimination in, 64
    litter box in, 64–65

B-complex vitamins, to increase, 27–28
Bed, for cat, 69
Behavior of cats, 141–143
  modification of, 59–66, 89
  problems, clues to:
    aggressive behavior, 22–23
    biting when handled, 17
    depression, 20
    and health of cat, 11–32
    hiding or irrational shyness, 18–19
    incessant crying, 24
    indiscriminate defecation, 14
    indiscriminate urination, 15
Belligerence:
  fear-induced, 38–42
  intermale, 47–50
  play, 50–53
  territorial, 42–47
Bill of Rights:
  for cat, 125–126
  for person, 126
Birth of baby, 55
Biting, when handled, 17–18
Bladder infections, 15, 16
Bladder stones, 17, 82
Blinking of eyes, 134
Blocked anal glands, 14
Blocked urinary tract, 16, 17
Blood in urine, 15, 16
Blows from paws, 135
Blue-eyed white cats, 150
Boarding of cats, 149
Body language, of person, 139–140
Bones, feeding of, 52–53
Book of Cats, Ross, 85
Books on cats, 3–4
Bootsie Mittens, Harvard Club cat, 120–121
Boredom:
  expressed by licking, 131
  remedies for, 72
Boxing stance, 43
Brainwashing, of fearful cat, 39–40
Breast cancer, 48

Breath, foul, 18
Breeds of cats, trainable, 151
Brewer's yeast, 28, 85
Bribes, 35
Burying of food, 79–80
Bushy tail, 133
Butterscotch, diabetic cat, 13

Cages for cats, 149
Calculi formation, 82
Calling of cat, 58, 92–99
Carroll, Lewis, *Through the Looking Glass,* 129
Carrying case, adaptation to, 109–111
Castration, 47–48
Cat expletive, 42
Cat flap, use of, 108–109
Catnip, 76
Cat owners, 3
Cat psychologists, 10
Cats, as wild creatures, 4
Cat's Bill of Rights, 125–126
Cat sitters, 146–149
The Cat's Pajamas, Fleischer, 85, 146
Catteries, 149
Cheeks, rubbing with, 54
Chemical treatment, for spraying problems, 60
Chewing on lamp cords, 78
China, cats in, 7
Chirring, 138
Choice of cats, 150–151
Clawing, 73
Cleanliness of litter boxes, 61–62
Clipping of claws, 104–105
Clouded eyes, 134
Coast Guard mascot, 11–12
Coat, dull, 23
Cod-liver oil, 85
Collars, 70
Cologne, cat sprayed with, 47
Come when called, 58, 92–99
Commercial cat food, 26, 81–83
  for anemia, 28
  ash content, 30

Commercial cat food *(cont'd)*
  for constipation, 28
  low-magnesium diet, 31
Communication with cats, 127–140
Conditioning of kittens, and
  trainability, 151
Consistency, in training, 52
Constipation, 24
  diet for, 28–29
  dry food for, 82
Contented licks, 132
Contents of litter box, 60, 61
Conversation with cats, 128
Cornucopia brand cat food, 26, 83
Covered litter boxes, 61
Crouching posture, 39, 42, 136
Crying, excessive, 24, 67–72
Cystitis, 15, 16, 82
  and misplaced elimination, 60

Danger, avoidance of, 119–122
Dangerous plants, 77
Dead animals, as gifts, 141
Deaf cats, 150
Death in family, 55
Declawed cats, 92
Declawing, 73–74, 88
Decreased urine volume, 15
Defecation, indiscriminate, 14–15
Deodorant spray, perfumed, 62
Depo-Provera, 60
Depression, 19–22
Depth of litter box, 61
Designer summons, 93–99
Destructive behavior, 72–79
Diabetes, 13, 15, 17
  and obesity, 22
Diarrhea, 14, 15, 23
  dry food for, 82
Diet, 25–32
  and bladder infections, 16
  new, 79
  problems of, 22
  variation in, 81–82
  vitamin additives, 21

Disease, and behavior problems, 89
Displacement attacks, 142
Docility, memory set, 37–38
Doctors, 143–145
Dogs, differences from cats, 6, 37, 124
Domestic cats, and wild cats, 6, 7
Double-faced Scotch tape:
  to discourage spraying, 58
  to protect house plants, 77
Dripping faucet, 142
Drooling, 18
Drooping tail, 132
Drooping whiskers, 137
Dry cat food, 82
  ash content of, 30
Dry fur, 23
Dry nose, 18
Dull coat, 23

Ears, position of, 133–134
  flattened, 39
Eating habits, picky, 79–85
Egypt, cats in, 7
Electric blankets, 70
Electricity, static, 17
Elimination procedure, 59
Embarrassment, expression of, 130
Entry flaps, 108–109
Environment, manipulation of, 88
Environmental requirements of
  cats, 6
Exercise, 69–70, 115–119
  for constipation, 29
Expectations of cats, 2
Eye contact, 139
Eye membranes, grayish, 20
Eyes, expression of, 134
  dilated, 39
  glazed, 18

Face hold, 51
Face-saving licks, 131
Falls, 121–122

Fasting, 79, 84
Fat Cat, Coast Guard mascot, 11–12
Fearful person, cat's reaction to, 56
Fear-induced belligerence, 38–42
Fecal analysis, 15
Feeding place, and spraying behavior, 57
Feeding practices, 79–85
  and coming when called, 94
  hand feeding, 35, 81, 96
Feline urologic syndrome (FUS), 12
  diet for, 29–31
Felix brand scratching post, 75
Female cats, spaying of, 48–49
Fetch, 118–119
Fever, 18–19, 20
Fiberglass cat carriers, 110
Fighting cats, 22, 47–50, 143
Finicky eating habits, 79–85
Fishing-line for leash, 106
Flattened ears, 133–134
Flatulence, 15
Fleischer, Leonore, *The Cat's Pajamas,* 85, 146
Food:
  availability of, 84
  requirements, 25–32
  as reward, 35
  as training reinforcement, 92
Footpads, sweating from, 18
Frequency of feeding, 85
Frightened cat, 120
  body posture of, 135
  coming when called, 97
  posture of, 39
  tail position, 133
Frontal stance, 135–136
Fur, dry, 23
Furniture, jumping on, 77–78
FUS (feline urologic syndrome), 12, 29–31

Gallico, Paul, 80
Garden, for cats, 76–77
Garlic odor for cat food, 85
Gastrointestinal disorders, 15

Genetic traits, 150
Gifts, dead bird or mouse, 141
Glazed eyes, 18
Golf balls, 143
Grief, expression of, 55–56
Grooming:
  excessive, 71
  surface conditions for, 42
Grooming licks, 131
Growling, 138
Guide to Nutritional Management of Small Animals, 26
Gums, pale, 20

Hairballs, 15, 71
Hair impaction, 29
Half-closed eyes, 134
Hand feeding, 81, 96
  and training, 35
Handling of cat, 36
Handling of kittens, 152
Harness:
  for outdoor play, 70
  for walking, 105–108
Harvard Club cat mascot, 120–121
Head rubbing, 132
Health of cat, 88–89
  and behavioral problems, 11–32
Hearing of cat, 93
Heart disease, 19
Heat, sources of, 70
Hematocrit value, 27
Hereditary characteristics, 150
Hiding, 18–19
Hiding places, 68–69
  cat carriers as, 109–111
High-rise syndrome, 121–122
High tail, 132
Hissing, 29, 138
Hissing at cat, 42
Hitting of cat, 37, 40
Holistic cat foods, 26
Home cooked cat food, 26–32
Home remedies:
  aggressive behavior, 23
  biting when handled, 17
  depression, 20

Home remedies *(cont'd)*
  fever, 20
  incessant crying, 24
  indiscriminate defecation, 14
  indiscriminate urination, 16
  loneliness, 68
  poor appetite, 81–85
  spraying, 57
Honesty of doctor, 145
Hormonal problems, 22, 23
Hormone treatments, 50
Housebreaking, problems of, 53–67
House plants, eating of, 76–77
How to Talk to Your Cat, Moyes, 80
Hugging, with paws, 135
Hunting instincts, 141–142
Hyperthyroid problems, 23
Hypoallergenic cat, diet for, 31–32
Hypothyroidism, 19

IAMS cat food, 82
Ill health, and misbehavior, 11–32
Illness, symptoms of, 18
Incessant crying, 24
Indiscriminate defecation, 14–15
Indiscriminate spraying, 22
Indiscriminate urination, 15–17, 56–66
Infants, human, development of, 8–9
Injury, pain from, 24
Instant action, for problem behavior, 56
Insulin injections, 17
Inter-cat fighting, 22, 47–50, 143
Introduction of new cat, 45–47, 55
Iron, in diet, 21
  to increase, 28
Isolation, to teach litter box behavior, 64

Japan, cats in, 7
Jealousy, and behavior problems, 56

Jean Naté After Bath Lotion, 78
Jumping on furniture, 77–78
Jump over a stick, 117
Jump up, 115–116

Kaolin, 14
Kaopectate, 14
Kittens, 152
  feeding of, 81
  orphaned, diet for, 30–31
  trainability of, 151
  training of, 34–36, 54
  teething, 52
Kneading, 71, 135
Kyolic, 85

Lamp cords, chewing on, 78
Language of cats, 127–140
Lashing tail, 131–132
Learning process of cats, 5–6
Leash, walking on, 105–108
Lewis, Lon D., 26
Licking, 130–132
Life span, and indoor living, 101
Limitations of training, 91
Liquid medication, 103–104
Listlessness, 19–22
Litter box:
  alternate materials for, 60
  cleanliness of, 57
  conditions of, 59–60, 61–63
  and spraying behavior, 57–59
Liver, raw, in diet, 28
Liver ailments, diet for, 32
Location of litter box, 62–63, 64
Loneliness, 68
  remedies for, 72
Longhaired cats, temperament of, 150
Loss, emotional, 19
  and house training problems, 55–56
Loss of appetite, 18
Lost cat, calling of, 98–99
Low ash cat foods, 30, 82–83
Lowered crouch, 136
Low-magnesium diet, 17, 28, 30
Lying on back, 136

Magnesium in diet, 17, 28, 30,
82–83
Male cat, neutering of, 47–48
Mammary gland tumors, 48
Marking of frightened people, 56
Marking of territory, 54, 56–66
Mark Morris Associates, 26–27
Mating calls, 139
Medical problems, 14–24
Medical solutions:
aggressive behavior, 23
illnesses, 19
indiscriminate defecation, 15
indiscriminate urination, 15
spraying problems, 60
Medicating of cat, 103–104
conditions for, 42
Memory, set of, 35
Memory enforcers, 58
Meow, silent, 80
Meowing, 138
Metal cat carriers, 110
Milk, for orphaned kitten, 30–31
Milk of bismuth, 14
Misbehavior, and ill health, 11–
32
Moist cat food, 26
Morris, Mark L., Jr., 26
Mothballs, 122
Mother cat, removal of kittens
from, 152
Mousetrap, upside-down, 79
Moyes, Patricia, *How to Talk to
Your Cat,* 80
Mucous membranes, pale, 21

Nagging body language, 136–137
Nagging paws, 135
Name of cat, 93
Nape of neck, lifting by, 51
Nasal passages, blocked, 81
Natural cat foods, 83
Negative reinforcement, 88
to coming when called, 94–95
Nervousness, licking of, 131
Neutering, 23, 47–48
and spraying, 56–57

New cat:
choice of, 143, 150–151
introduction of, 45–47, 55, 68
New foods, introduction of, 83
Nighttime crying, 71
"No" command, 34, 40, 45, 91
discipline with, 56
memory set with, 51
Noise therapy, 102
for spraying, 59
Nose:
communication with, 137
dry, 18
pale, 20
Nutrition, 22, 25–32

Obesity, 21, 22
Odor of food, 85
Odors, to remove, 63
Old age, problems of, 15
Onion powder, 85
Orange peels, to protect
furniture, 78
Orphaned kitten, diet for, 30–31
Outdoor cat:
living indoors, 101–102
declawed, 74
Overcrowding, 143
Overgrooming, 131
Overweight, 19–22
Owners of cats, 3

Painful associations with litter
box, 60, 63–64
Palpation, 18, 24
Parasitic cause of anemia, 21
Par-for-the-course theory, 89–91
Paws, communication with, 135
Pebbles, decorative, to protect
house plants, 77
Pectin, 14
People-watching, 72
Pepper:
to discourage chewing on
lamp cords, 78
to protect house plants, 77

Permission seeking, by licking, 130
Persistence in training, 89–90
Personality problems, 33
  crybaby, 67–72
  destructive behavior, 72–79
  killer cat, 33–53
  prima donnas, 79–86
  uncouth cat, 53–67
Person's Bill of Rights, 126
Petroleum jelly, 29
Pets, healing qualities of, 86
Physical punishment, 37, 40
Physical state of cat, altering of, 88
Pills, administration of, 103
Ping-Pong balls, 69
Plant atomizer, *see* Water gun
Plants, eating of, 76–77
Play:
  belligerent, 50–53
  for exercise, 69–70
  rough, between cats, 46
Pointed ears, 133
Poisonous plants, 77
Poisonous substances, 122
Population density, 43, 143
Positive reinforcement, 42, 88, 91–92, 115
  of kitten, 34
Posture of cat, 135–136
  of fearful cat, 39
Practice, 124
Praise:
  for good behavior, 76
  as training tool, 92, 100
Privacy, need for, 68–69
Private bathroom for cats, 114–115
Private places, cat carriers as, 109–111
Progestins, for spraying behavior, 60
Property rules, 141
Protein in diet, to increase, 27
Psychology of cats, 141–143
Punctuality of feeding, 85
Punishment, physical, 37, 40

Pupils of eyes, narrowed, 134
Purring, 137–138

Rectal thermometer, to use, 19
Regularity of feeding, 85
Release command, 99–100
Renal ailments, 17
  diet for, 32
Repetition in training, 90, 100–101, 124
Rewards, 52
  for coming when called, 96
  food as, 35
Rhubarb, seeing-eye cat, 9
Ring for out, 117
Ritual, 95
Rolling over, 123
Rope, safety precautions, 57
Ross, Charles Henry, *Book of Cats,* 85
Rough play, 37–38
  between cats, 46
Routine, 95
Runt of the litter, 151

Sadness at perceived loss, and loss of house training, 55–56
Salt, to protect house plants, 77
Salting of food, 16
Schneider, Mrs. Elsie, 9
Scratching, 73
  around dish, 79–80
Scratching posts, 74–76
Screens, to protect house plants, 77
Scruff of neck, lifting by, 51
Sebaceous glands, overactive, 14
Seeing-eye cat, 9
Self-fulfilling prophecies, 2
Separation, 19
Set the memory, 35
Shaking of cat, 51
Sharing of territory, 44–47
Shelves, forbidden, 78–79
Shivering, 18
Shnoogling, 35–36
Shock reaction, 18

Shortness of breath, 20
Shyness, irrational, 18–19
Sideways stance, 135
Silent meow, 80
Silver, stray cat, 12–13
Sit up, 116–117
Skin lesions, 23
Skunk deodorant, tomato juice
   as, 111
Sleeping arrangement, 69
Smell, sense of, 81
Smooth surfaces, fearful
   response to, 42
Social stress, 15
   and health, 142
   and indiscriminate urination,
   57
Soft-moist cat foods, ash
   content, 30
Spaying of female cats, 48–49
Spitting, 39
Spitting at cat, 42
Spoilage of cat food, 83
Spoiled behavior, 79–85
Spray gun, *see* Water gun
Spraying, 56–66
   indiscriminate, 22
Staring eyes, 134
Start doing something, training
   techniques, 89–125
Static electricity, 17, 18
Stay command, 99–101
Stimulus, new, adaptation to, 142
Stop command, 99–101
Stop doing something, training
   techniques, 33–86, 88–89
Street cats, 150
Stress, 14
   and health, 142
   social, 15
     and indiscriminate
     urination, 57
Striking of cat, 37, 40
Stud tail, 14, 113–114
Substrate, 6–7, 59
Sucking, 71–72
Summons of cat, 93–99
Sweating from footpads, 18

Tail:
   cleaning of, 113–114
   position of, 132–133
   twitching, 129
   waving, 39
Temperature:
   high, *see* Fever
   to take, 19
Territorial belligerence, 42–47
Territorial marking, 54
Territory claims, 141
Thermometer, to use, 19
Thirst, increased, 13, 15
Through the Looking Glass,
   Carroll, 129
Thyroid problems, 22, 23
Time of feeding, 85
TLC, 55–56
Toilet training, 6–7, 65–66
Tomato juice, as skunk
   deodorant, 111
Tongue licks, 130–132
Touch, 36
   and affection, 51
Toys, 142
Trace minerals, in diet, 28
Training of cats, 1–2, 7–10, 87–89
   coming when called, 96–97
   and licking behavior, 130–132
   and tail position, 132–133
Tranquilizers, 40, 49
Treats, feeding of, 80–81, 96
Tricks, teaching of, 88, 115–119
Tumors, 15, 17, 18
Tuna, diet of, 25
Twitching tail, 132–133
Twitching whiskers, 137
Tylenol, 19, 20

Unneutered cats, urine marking
   by, 56–57
Upright ears, 133
Upside-down mousetrap, 79
Urinary blockage, 16, 24, 82
   diet for, 29–30
Urinary calculi, 12
Urinary incontinence, 15

Urination, indiscriminate, 15–17, 56–66
Urine:
  blood in, 15, 16
  decreased volume, 16
Urine marking, 56–66
Uterine infections, 49

Valium, 49
Veterinarians, 143–145
  author's training as, 4–5
Vinegar:
  to discourage chewing on lamp cords, 78
  to discourage misplaced elimination, 62, 63
  to discourage spraying, 58
Vitamin B complex, 21
Vitamin C, for cystitis, 16
Vitamin E, deficiency of, 25
Vitamins, overuse of, 22, 26

Walking on a leash, 105–108
Water, for drinking, 82, 84
Water guns, 41, 50, 56
  for destructive behavior, 75–76
Wax paper, to discourage spraying, 58
Weaning, early, 71
Weight loss, 23
  as remedy for arthritis, 16

Whiskers, communication by, 137
Whistles, 93, 102
White, Diane, "You Know How Cats Are," 153–155
White cats, blue-eyed, 150
Wicker cat carriers, 110
Wide open eyes, 134
Wild cats, 5
  differences from domestic cats, 7
  evacuation habits of, 54
  similarities to domestic cats, 6
Wild creatures, cats as, 4
Windows, falls from, 121–122
Windowsills, 72
Wire, safety precautions, 57
Wool, sucking of, 71–72
Workouts, 115–119
Wounds, 17, 18
Writers of cat books, 3–4

X-rays, 145

"You Know How Cats Are," White, 153–155
Yowling, 138

Zoos:
  animals in, 5
  big cats in, 53